Millennials, Generation Z and the Future of Tourism

THE FUTURE OF TOURISM

Series Editors: **Ian Yeoman**, *Victoria University of Wellington, New Zealand* and **Una McMahon-Beattie**, *Ulster University, Northern Ireland, UK*

Some would say that the only certainties are birth and death; everything else that happens in between is uncertain. Uncertainty stems from risk, a lack of understanding or a lack of familiarity. Whether it is political instability, autonomous transport, hypersonic travel or peak oil, the future of tourism is full of uncertainty but it can be explained or imagined through trend analysis, economic forecasting or scenario planning.

This new book series, The Future of Tourism, sets out to address the challenges and unexplained futures of tourism, events and hospitality. By addressing the big questions of change, examining new theories and frameworks or critical issues pertaining to research or industry, the series will stretch your understanding and generate dialogue about the future. By adopting a multidisciplinary perspective, be it through science fiction or computer-generated equilibrium modelling of tourism economies, the series will explain and structure the future – to help researchers, managers and students understand how futures could occur. The series welcomes proposals on emerging trends and critical issues across the tourism industry and research. All proposals must emphasise the future and be embedded in research.

All books in this series are externally peer-reviewed.

Full details of all the books in this series and of all our other publications can be found on http://www.channelviewpublications.com, or by writing to Channel View Publications, St Nicholas House, 31-34 High Street, Bristol BS1 2AW, UK.

THE FUTURE OF TOURISM: 7

Millennials, Generation Z and the Future of Tourism

Fabio Corbisiero,
Salvatore Monaco and
Elisabetta Ruspini

CHANNEL VIEW PUBLICATIONS
Bristol • Jackson

DOI https://doi.org/10.21832/CORBIS7611
Library of Congress Cataloging in Publication Data
A catalog record for this book is available from the Library of Congress.
Names: Corbisiero, Fabio, author. | Monaco, Salvatore, author. | Ruspini, Elisabetta, author.
Title: Millennials, Generation Z and the Future of Tourism/ Fabio Corbisiero, Salvatore Monaco, and Elisabetta Ruspini.
Description: Jackson, TN: Channel View Publications, [2022] | Series: The Future of Tourism: 7 | Includes bibliographical references and index. | Summary: "This book examines the lifestyles, expectations and plans of Millennials and Generation Z and how they are redefining tourism. It explores the present and future challenges faced by the tourism industry as a result of the generational turnover and the role a generational perspective can play in helping the industry recover from the COVID-19 crisis"—Provided by publisher.
Identifiers: LCCN 2021060701 (print) | LCCN 2021060702 (ebook) | ISBN 9781845417611 (Hardback) | ISBN 9781845417604 (Paperback) | ISBN 9781845417628 (PDF) | ISBN 9781845417635 (ePub)
Subjects: LCSH: Tourism—Social aspects. | Hospitality industry—Social aspects. | Leisure industry—Social aspects. | Generation Y—Attitudes. | Generation Z—Attitudes. | Popular culture—Social aspects.
Classification: LCC G156.5.S63 C67 2022 (print) | LCC G156.5.S63 (ebook) | DDC 306.4/819—dc23/eng20220311
LC record available at https://lccn.loc.gov/2021060701
LC ebook record available at https://lccn.loc.gov/2021060702

British Librar1y Cataloguing in Publication Data
A catalogue entry for this book is available from the British Library.

ISBN-13: 978-1-84541-761-1 (hbk)
ISBN-13: 978-1-84541-760-4 (pbk)

Channel View Publications
UK: St Nicholas House, 31-34 High Street, Bristol, BS1 2AW, UK.
USA: Ingram, Jackson, TN, USA.

Website: www.channelviewpublications.com
Twitter: Channel_View
Facebook: https://www.facebook.com/channelviewpublications
Blog: www.channelviewpublications.wordpress.com

Copyright © 2022 Fabio Corbisiero, Salvatore Monaco and Elisabetta Ruspini.

All rights reserved. No part of this work may be reproduced in any form or by any means without permission in writing from the publisher.

The policy of Multilingual Matters/Channel View Publications is to use papers that are natural, renewable and recyclable products, made from wood grown in sustainable forests. In the manufacturing process of our books, and to further support our policy, preference is given to printers that have FSC and PEFC Chain of Custody certification. The FSC and/or PEFC logos will appear on those books where full certification has been granted to the printer concerned.

Typeset by Deanta Global Publishing Services, Chennai, India.

Contents

Introduction: Tourism at the Crossroads: Between Past,
Present and Future   vii
*Fabio Corbisiero, Salvatore Monaco and
Elisabetta Ruspini*

Part 1: On the Go with the New Generations

1   Generations, Events, Experiences, Tourism   3
    Elisabetta Ruspini

    Introduction   3
    Generational Theories   4
    Generations and Tourism Preferences   13
    Conclusion   22

2   Capturing the Future Traveller   26
    Salvatore Monaco

    Introduction   26
    Being Young Today: A Social Identikit   27
    Main Consumer Choices of New Generations   29
    The Role of Social Media and Influencers   33
    Tourist Choices and Instagrammability   36
    In Search of Sustainability   38
    Conclusion   41

Part 2: Technologies and the Sharing Economy in Tourism

3   NetGen and Tourism   47
    Fabio Corbisiero

    Introduction   47
    Technological Advances in Tourism   50
    The Hyper-Connected Generations   53
    Conclusion   60

4 Towards a Sharing Generation *Salvatore Monaco*	63
Introduction	63
Sharing Car and Home	64
Sharing to Save	71
The Sustainable Approach to Consumption	73
Tourism as a Life Experience	75
The Shadows of Sharing	78
Conclusion	80

Part 3: Generations, Gender and LGBT Issues in Tourism

5 Gender, Generations, Tourism *Elisabetta Ruspini*	85
Introduction	85
The Past	87
Towards the Present	90
Between Present and Future: Millennials and Gen Zers	93
Past, Present, Visions of the Future	97
6 LGBTQ+ and Next-Gen Tourism *Fabio Corbisiero*	103
Introduction	103
From Grand Tour to the Present: The Rainbow Travel	106
What Kind of Travellers are LGBTQ+ People?	109
The Generation Lens to Observe LGBTQ+ Tourism	111
Conclusion	117
Conclusion: Understanding the Present to Prepare for the Future (of Tourism) *Fabio Corbisiero, Salvatore Monaco* *and Elisabetta Ruspini*	120
References	126
Index	154

Introduction

Tourism at the Crossroads: Between Past, Present and Future

Fabio Corbisiero, Salvatore Monaco and Elisabetta Ruspini

The book aims to highlight the present and future challenges posed by the generational turnover to tourism. The concept of generation is one of the most important sociological lens to interpret social change (Ariès, 1979), a key concept to understand the tension between continuity and social transformations, between past, present and future. As Karl Mannheim (1928) suggests, a generational location limits and predisposes individuals to specific behaviours and feelings. People born at the same time are similarly exposed to primary and secondary socialisation and grow up under the influence of the same historical events. Each generation develops unique needs, expectations, forms of sociality, cultural and communicative practices and consumption choices. In this sense, each generation is different from the previous and subsequent ones due to its own peculiar characteristics, values and visions of life. Studies of social change (economic, political and cultural transformations, scientific and technological advances, value transmission, social mobility, family choices, employment trends, lifestyles) must all cope with generational peculiarities and relations (Kertzer, 1983). The relationship between tourism and generations is also dynamic and, as such, should be carefully considered and monitored. As Yeoman and McMahon-Beattie (2020) argue, the current and potential future existence of the tourism industry is influenced by the link between past and present. Thus, in order to ensure a strong, inclusive and sustainable future for tourism, the tourism industry should pay specific attention to the generational turnover and to the peculiarities of the younger generations. Tourism literature studies that use a generational perspective are still limited: in recent years, however, there has been growing attention on generational analysis, emphasising significant differences in tourist behaviours among generations.

In light of the above, the goal of this book is to focus on lifestyles, expectations and plans of the younger generations (from the millennial generation onwards) and how these generational groups are redefining and will redefine tourism. Putting younger travellers at the centre of the scientific reflection seems a useful approach not only for identifying current tourist trends, but also for capturing the most significant peculiarities that will shape the future of tourism, whatever it will be. Young people link both past, present and future and anticipate social trends. On the one hand, they inherit values and habits from previous generations. On the other hand, their culture differs from the culture of older generations and they have the power to shape the world of tomorrow. One of the main ideas behind our reasoning is precisely that an understanding of generational shifts in tourist behaviour can help the identification of future tourism trends (Corbisiero & Ruspini, 2018). This idea is in line with the theoretical assumptions of the sociology of the future, where prediction can be accomplished by extending and amplifying present trends (e.g. Bell, 2009; Jacobsen & Tester, 2012; Wagar, 1983). Tourism forecasting is a key tool to identify future trends and to test possible tourism scenarios that may affect the tourism industry at a national or global level, involving multiple stakeholders and tourism professionals. Notwithstanding this, thinking about the future of tourism has been under researched (Yeoman, 2008, 2012).

A number of questions are addressed by the chapters contained in this book: How is generational turnover affecting tourism choices? What is the role of new technologies and social networks in the tourist experience? What are the challenges for tourism in the near future? Can we identify them, using a generational perspective? Can a generational perspective be useful to help the tourism industry recover from the Covid-19 crisis? The generations of travellers considered are mainly millennials and Generation Zers. However, the very young members of Generation Alpha are also taken into account.

From a theoretical point of view, we will refer to different generational theories inspired by Mannheim's seminal work, such as the one proposed by Neil Howe and William Strauss (1991), who have long studied generational dynamics and generational change. The two authors identify different generations in the history of the US, attributing specific characteristics to each of them. We know that no clear dividing line separates one generation from another. However, Strauss and Howe's analytical categories are a useful tool for understanding values, desires and forms of consumption that are common to people who are born and live in the same period of time.

The millennial generation (currently the youngest adult generation, women and men whose birth years roughly range from 1982 to the late 1990s) has been defined as one that is diverse, technological, connected and open to change (Benckendorff et al., 2010; Howe & Strauss, 2008; Taylor & Keeter, 2010; Rainer & Rainer, 2011). Millennials are one of the most culturally and ethnically diverse generations in history. They are IT literate (they grew up with 2.0 web technologies) and open to new family arrangements and gender equality. Millennials are also closely interconnected with their families and friends. Compared with their elders today, millennials are much less likely to affiliate with any religious tradition. Because the economic and financial crisis of the last years has deeply affected this generation and its lifestyle, millennials are increasingly price-sensitive and attentive to shared-economy activities: they are concerned about economic sustainability and sustainable development (Ranzini et al., 2015; Bernardi & Ruspini, 2018; Deloitte, 2021). In regard to tourism, millennials are a lot more willing to spend money on experiences that enhance their lives than on material things. Following the study on 'Tourism Megatrends' by Horwath HTL (2015), this rapidly growing segment is expected to represent 50% of all travellers by 2025.

Generation Z is a common name for the group of people born roughly from the end of the 1990s through the early 2010s, a span of 15–20 years in the very late 20th and very early 21st centuries. Gen Z members are digital natives: they have grown up in an era of mobile devices and smartphones and have been raised in a world of social websites. They are multitaskers and prefer media that are simple to use and interactive as opposed to 'passive' TV. Technology has given these young people an unprecedented degree of connectivity among themselves and with the rest of the population. This generation is entrepreneurial, innovative and concerned with effecting social change (Seemiller & Grace, 2016). Gen Z members have global values and are environmentally conscious consumers (Deloitte, 2021). Gen Zers like a more diverse range of holidays compared to previous generations, expect to access and evaluate a broad range of information before purchasing (Wood, 2013) and give great importance to experiential tourism.

Generation Alpha children were born between 2010 and 2015. Also known as the 'children of millennials', they are the first generation entirely born within the 21st century. This generation is considered by many to be the more tech-savvy generation to date. Thus, one key question is how emerging technologies, such as artificial intelligence (AI), will impact their behaviours, choices and lifestyles.

The book is divided into three parts.

Part 1: On the Go with the New Generations includes two chapters aimed at offering a general overview of the importance of applying a generational approach to the study of tourism and defining the profile of young world travellers. This first section of the book discusses, on the one hand, the value of the concept of social generations for tourism studies and provides readers with the theoretical tools to understand the generational approach. On the other hand, it offers an identikit of today's travellers, paying attention not only to differences among generational groups, but also between the same generations on a territorial basis. Borrowing from both Mannheim's work and other prominent theories of generations, Chapter 1: *Generations, Events, Experiences, Tourism*, clarifies what a generational theory is. It also deepens the relationship between generations and tourism choices, exploring current applications of the generational concept in tourism. Specific attention is paid to the examination of the main features of millennials and Gen Z members, underlying the centrality of these social groups for tourism studies. Chapter 2: *Capturing the Future Traveller*, is aimed at defining the profile of millennials and Generation Z members as both consumers and travellers, starting from relevant studies that have been conducted both locally and globally. Specific attention is devoted to the relationship between millennials/Gen Zers, tourism and social networks, with particular reference to social networks based on images, such as Instagram. The online collaborative consumption platforms are generating a series of new practices that give new meaning to the travel experience.

Technology drives the second part of the book: Technologies and the Sharing Economy in Tourism. Chapter 3: *NetGen and Tourism*, points out the advent of new technologies and its tie with millennials and Gen Zers. Technology has generated a series of mutations in the dynamics and structure of production and consumption at both global and local levels and this impacts on tourism choices and needs. The chapter first highlights the differences in the use of digital technologies between millennials and other generations (baby boomers, Gen Zers, etc.). It then discusses how digital technologies have influenced the travel behaviour of millennials and other generations, a very relevant issue for the travel and tourism industry to implement a successful marketing strategy. A meta-analysis approach is used, taking different studies into account that focuses on tourism and marketing concepts in the era of information technology. How is technology changing and influencing the link between tourism and the new generations? How do digital technologies impact young tourists' consumption experiences? Chapter 4: *Towards a Sharing Generation*, discusses the link between tourism, generations and a new form of reciprocity made possible by the use of sharing platforms. The sharing economy has emerged in response to the economic crisis of 2008–2011 and has many relationships with the tourism sector (European Parliament, 2015; Finger *et al.*, 2017). Technologies and the use of online collaborative consumption platforms are accentuating and consolidating a (post)modern version of reciprocity, an economic model no longer centred around ownership but on the sharing of things, less based on the purchase of goods and services and more on their use (Smith, 2016). In particular, several studies (e.g. Barbosa & Fonseca, 2019; Deloitte, 2015; Fondevila-Gascón *et al.*, 2019; Ianole-Calin *et al.*, 2020) have highlighted how the younger generations are implementing new forms of consumption, defined in the literature as 'collaborative'. The chapter critically analyses the most common sharing practices that characterise the tourism sector, such as sharing means of transport, holiday home exchange, couchsurfing and team purchase.

Part 3: Generations, Gender and LGBTQ+ Issues in Tourism includes the last two chapters, designed with the aim of highlighting forms of social exclusion in the tourism sector (gender inequalities and discriminations linked to sexual orientation) and understanding how the new generations are facing these challenges. Chapter 5: *Gender, Generations, Tourism*, incorporates a gender perspective into the generational analysis. The chapter first offers a brief overview of the history of women's travel and constraints on women's access to leisure. It then analyses some of the peculiarities of millennial and Gen Z travellers, addressing the changes in gender roles and values. It then discusses how the outbreak of the coronavirus disease (Covid-19) may reverse positive trends connected to the increasing women's participation in leisure and tourism. The unknown evolution of the pandemic and the impact of the resulting restrictions imposed to fight the spread of the disease present unprecedented challenges to both the tourism sector and gender equality (UN, 2020a, 2020b; UNWTO, 2020a, 2020b). The focus of Chapter 6: *LGBTQ+ and Next-Gen Tourism* is on tourism practiced by gay, lesbian, bisexual, transgender and queer people. Even though the study of LGBTQ+ generations in the tourism market is fairly recent, it represents a key issue in the late-modern development of tourism research, strongly linked to the generational turnover. The rate of LGBT+ identification is indeed increasing among the younger generations (GenForward, 2018). The tourism industry is now called upon to focus on this burgeoning segment (Hughes, 2006): today, the key to success is not mere mass-marketing but rather a strong 'niche position' that offers something singular to a particular set of consumers. This chapter also focuses on the relationship between gender, sexual orientation and tourism, discussing 'rainbow turistification' (Corbisiero, 2014). The chapter concludes by saying that when progress is made towards equal rights (e.g. the introduction of same-sex marriage), destinations benefit from a boost to their brand, and increased arrivals and spending associated with wedding ceremonies, receptions and honeymoons.

The concluding chapter summarises the main issues faced in the book and highlights the need to consider climate change, global warming and environmental concerns, along with the Covid-19 pandemic, in order to gain new insights into the future of tourism from a generational lens.

Part 1
On the Go with the New Generations

1 Generations, Events, Experiences, Tourism

Elisabetta Ruspini

Introduction

This introductory chapter offers a theoretical reflection on the usefulness of applying a generational approach to the study of tourism. The notion of generation is one of the most important sociological concepts to describe both the nature of the mutual relation between individual and society (Alwin & McCammon, 2007) and to explain social change. Generation is a measure of historical time. The term 'generation' refers to a group with a common and distinctive identity shaped by experience, and the stratification of this experience, through time. Generation has been defined (Gilleard, 2004) as a temporally located 'cultural field' emerging at a particular moment in history within which individuals from a potential variety of overlapping birth cohorts participate as generational agents. Exposure to key historical events that took place during each cohort's transition to adulthood provides the markers for each generational field (Gilleard, 2004: 112, 114). Thus, generation covers a wide range of cohorts. Generational theory emanates from Karl Mannheim's (1928, 1952) fundamental work on generations, identity and knowledge, which suggests that cultural differences between generations come from significant (social, economic or political) events which have occurred during formative years, from being exposed to analogous primary and secondary socialisation processes, and from the interchange between young people and society's accumulated cultural heritage. The formative experiences of each generation influence its members' lifelong beliefs, values and behaviour: each generation has peculiar characteristics, needs and expectations, unique to that generational group.

In the tourism literature, there are still limited studies of generational change. However, recent years have witnessed an increasing interest in generational analysis in tourism studies, highlighting differences in tourist behaviours among generations (Beldona *et al.*, 2009; Benckendorff *et al.*, 2010; Chiang *et al.*, 2014; Corbisiero, 2020; Corbisiero & Ruspini, 2018; Haydam *et al.*, 2017; Huang & Lu, 2017; Li *et al.*, 2013; Pennington-Gray *et al.*, 2002; Southan, 2017). Those and other studies have shown that knowledge gained through the lens of generational theory can provide useful theoretical insight and practical implications for tourism scholars and practitioners (Pendergast, 2010). Research has not only shown the value of generational analysis in tourism research, but has also concluded that a greater use of generational analysis is needed to examine changes in travel behaviour (Gardiner *et al.*, 2014; Li *et al.*, 2013; Oppermann, 1995). Moreover, as written by Hansen and Leuty (2012: 34), organisations today are faced with the challenges of integrating different generations in the workplace as well as the complexity of creating environments to attract and satisfy workers of each generation.

In light of the above, borrowing from both Mannheim's seminal work and other prominent theories of generations, this chapter explains what a generational theory is. It then discusses the relationship between generations and tourism choices, investigating contemporary applications of the generational concept in tourism. Specific attention is devoted to the analysis of the main characteristics of millennials and Gen Z members. Getting to know the younger generations is crucial since they represent the future of travel and tourism (Corbisiero & Ruspini, 2018). To conclude, the final section discusses the pros and cons of generational analysis.

Generational Theories

Generation is a rich and complex concept that encompasses academic disciplines ranging from biology to social anthropology, to psychology, to demography, to philosophy, to history and sociology (Jansen, 1974; Kertzer, 1983). The complexity of the concept of generation has posed difficulties for researchers in various areas. Its multidimensional nature has attracted a wide array of research methods and methodologies, often leading to paradigmatic conflicts and disagreements (Allen *et al.*, 2015). As Kertzer (1983) notes, in the social sciences different concepts of generations overlap. Kertzer (1983: 125–126) places them in four categories: generation as a principle of kinship descent; generation as cohort; generation as life stage; and generation as historical period. The concept of generation has a long tradition in social anthropology: social anthropologists use it in referring not only to the relationship between parents and their children, but also to the larger universe of kinship relations. The concept of generation has been used by anthropologists to explain social change over time, as it implies relations in a temporal perspective (Lamb, 2015). Demographers refer to a cohort notion of generations: people moving through the age strata, the younger replacing the older as all age together. According to Kertzer, Piotr Sorokin's (1947) discussion of conflicts between younger and older generations is a well-known example of its life-stage usage, while Shmuel Eisenstadt's (1956) classic study combined the descent and life-stage meanings of generation. Eisenstadt's

(1956: 9) basic hypothesis concerns the 'existence of age groups in societies not regulated by kinship or other particularistic criteria'. The use of generation to characterise the people living in a particular historical period is more common in history (Tannenbaum, 1976; Wohl, 1979): in this sense, generation covers a wide range of cohorts.

Generation as a social concept is a modern one. The definition of generation in sociohistorical terms emerged at the beginning of the 19th century. Views and theories concerning the generations of this period began to withdraw from the ancient meaning originating from the biological–genealogical framework (Kortti, 2011) — possibly because it was a time of accelerating historical change (Jaeger, 1985). As noted (Bristow, 2015a), industrialisation emphasised a conflict between continuity and change, and part of this was a disruption of stable generational boundaries: the industrial society, and its institutions, weakened the significance of kinship. As argued by Eisenstadt (1963), especially in the first phase of modernisation, there has been a growing discontinuity between the lives of children. New and enlarged perspectives in the social world favoured the development of generational tensions, particularly in relation to youth, and this challenged a model of life governed by kinship relations. This tendency brought out the need to use the concept of 'generation' as a framework in which to explain some of the sweeping changes that were happening. The concept of generation became a way of interpreting social change and organising thoughts about such radical change (Donati, 1995). Auguste Comte, John Stuart Mill and Wilhelm Dilthey were three of the most important thinkers to reflect on the generations of the 19th century and they created new viewpoints compared with previous studies. August Comte (1849) was the first to begin a scientific study of generations in history. Comte systematically examined the succession of generations as the moving force in historical progress. John Stuart Mill (1865) later expressed similar ideas by arguing that human beings are historically shaped and made what they are by the accumulated influence of past generations (Bouton, 1965). In his study of German romanticism, Wilhelm Dilthey (1875) observed that the absorption of formative impressions during adolescence tends to transmit for life to a great number of individuals of the same age relatively homogeneous philosophical, social and cultural guidelines (Jaeger, 1985: 275–276). Dilthey (1910) suggested that each epoch defines a life-horizon by which people orient their life, life concerns and life experience.

However, views from the 19th century did not yet include the idea of experience uniting generation members and the generational consciousness produced by it (Kortti, 2011: 70–71). Generations were still mainly regarded as age groups, even though as the century progressed the concept began to take on a cultural meaning, creating a framework in which to explain some of the sweeping changes happening. All of the most important modern generation theories were formulated in Europe starting from the 20th century, particularly in the 1920s. The group of generation theoreticians from the beginning of the 20th century includes the French cultural philosopher François Mentré (1920), the German art historian Wilhelm Pinder (1926), the Spanish philosopher and humanist José Ortega y Gasset (1933) and the German sociologist Karl Mannheim (1928, 1952). According to Woodman (2016: 20), the first sustained formalised sociological theorising of generations appeared along with the development of youth research. European scholars of generation (Mannheim, 1952; Mentré, 1920; Ortega y Gassett, 1933) were showing that youth could vary over time, with effects for the future of societies.

Contemporary generational theory focuses on the influence and impact of sociohistorical processes on the development of the shared meaning of events and experiences of age groups, highlighting the cultural aspects of cohorts. Generational theory is based on the concept that generations are different from each other due to significant social, economic or political events (especially if they occurred during the formative period, i.e. adolescence and early adulthood), resulting in specific and markedly different values, attitudes and lifestyles (Singer & Prideaux, 2006). People who are in young adulthood during significant national or international events will form a shared memory of those events which will affect their future attitudes, preferences and behaviour (Parry & Urwin, 2011: 81; Schuman & Scott, 1989). The conceptualisation of a generation as individuals sharing a common location in the historical dimension of the social process, youth experiences marked by unique social circumstances and events, and a fundamental sense of a common history, is rooted in Karl Mannheim's theory.

Karl Mannheim's work

Mannheim provided the most comprehensive and systematic treatment of the concept from a sociological viewpoint and modern empirical studies of generations proceed from his theoretical contributions. In Mannheim's (1927/28) classic treatise 'The Sociological Problem of Generations', he outlined the idea that generation is to be understood as a complex set of social interconnections situated in a given historical period. The social phenomenon of generations embraces related 'age groups' similarly embedded in history, in so far as they all are exposed to the same phase of the sociohistorical process (Spitzer, 1973). Mannheim pays particular attention to the dialectical relationship between the pace of change and the succession of generations, as well as to the potential contribution of young people to social and cultural change (Merico, 2012), emphasising the agency of youth. As noted (Woodman, 2016, 2018), Mannheim began to think in sociological terms about generational change in the wake of World War I and its impact on the young generation. He argued that ways of life pursued by the previous generation became at points in the historical process difficult to maintain, for example after a shared traumatic experience like the Great War. More specifically, his aim was to understand the way different cohorts of German youth contested the ideas inherited from their parents' generation, and how these groups could become the source of new values and new political movements. The 'problems of generations', from Mannheim's point of view, was to understand how ideas were transmitted and reshaped over time: in sociological terms, the construction and reconstruction of knowledge within multiple social forces (Bristow, 2015b).

Mannheim identifies three key dimensions: generation location ('Generationslagerung'), generation as an actuality ('Generationszusammenhang') and generation unit ('Generationseinheit') (Diepstraten et al., 1999; Simirenko, 1966). A 'generation location' refers to individuals who are located within the same generation by birth and are consequently exposed to a common range of events. By participating in the same historical and social circumstances, members of the same generation share, at least to a significant degree, experiences and challenges and belong to a common destiny in cultural and historical terms. The location in which individuals are socialised functions as a structure of opportunities which might be translated into a common culture or worldview. However, according to Mannheim (1952: 297): 'The fact that people are born at the same time, or that their youth, adulthood, and old age coincide, does not in itself involve similarity of location; what does create a similar location is that they are in a position to experience the same events and data, etc. and especially that these experiences impinge upon a similarly "stratified" consciousness'. In other words, not every generation location creates new collective impulses and formative principles and develops an original and distinctive consciousness (Mannheim, 1952: 309). There needs to be an active recognition of the shared experiences (Knight, 2009). 'Generation as an actuality' only arises where a concrete bond and generational consciousness are created between its members by their being exposed to the same social and intellectual stimulus. This conceptualisation presupposes that generation members subjectively identify with their generation (Diepstraten et al., 1999): a 'generation as an actuality' is constituted when people sharing a similar location in history also participate in a common destiny. Mannheim (1952: 304; Pilcher, 1994: 490) expresses the difference between 'generational location' and 'generation as actuality' as that of potentially being capable of actively participating in the 'characteristic social and intellectual currents of their society and period'. In Mannheim's theory, the step from a generation location to a generation as an actuality depends on the recognition of common experiences during the formative years, the years of youth (Pilcher, 1994). This is because generation members need to be old enough to deeply experience life-defining events while at the same time be in a young enough life stage where they can significantly influence their worldviews. According to Mannheim (1952: 304): 'Youth experiencing the same concrete historical problems may be said to be part of the same actual generation; while those groups within the same actual generation which work up the material of their common experiences in different specific ways, constitute separate generation units'.

The 'generation unit' represents a much more concrete bond than the actual generation as such because of the converging responses it involves. Formative and interpretive principles form a link between spatially separated individuals who may never come into personal contact at all (Mannheim, 1952: 306). Generation units express the generation style in their action: the German youth movement is Mannheim's example. Members of new generations can emerge as change agents both challenging traditional interpretations of historical conditions and offering an alternative interpretation, standing in opposition to previous generations and the cultural heritage they represent (Demartini, 1985: 2). Mannheim (1952: 293) emphasises the importance of 'fresh contacts' with the prevailing culture, that is, between young people and society's accumulated cultural heritage. The phenomenon of 'fresh contact' is a productive force: each new generation has the opportunity to reinterpret the existing cultural heritage and this ensures the continuous renewal of culture. Future engagements will build upon early cultural engagements and the capacity to rework the cultural legacy of previous generations (Woodman, 2018). If generation shapes subjectivity by both delimiting the range of possible beliefs and actions and providing the catalyst for new social movements and generational change within a culture (Mannheim, 1952: 303), this subjective dimension does not mean that all young people share the same beliefs or values. According to Mannheim, contemporaneous individuals are internally stratified by their physical location, culture, gender and class, and this stratification causes members to see the world differently. Mannheim explained that even if young people can develop a sense of belonging to the same generation oriented to analogous life experiences and challenges, differences and conflicts are possible (Woodman, 2016, 2018): members of a generation could have 'polar' responses to a shared generation location (Mannheim, 1952: 304; also Ortega y Gasset, 1933). On the one hand, each actual generation is subdivided into a number of generation units: different individuals, while experiencing common sociohistorical stimuli, may respond to them in a different manner. On the other hand, generation units build upon the consciousness of belonging to one generation.

A generation unit must contain two related and essential elements: a common location in time (generation location) and a distinct consciousness of that historical position, a 'mentalité' or 'entelechy'[1] shaped by the events and experiences of that time. In Mannheim's formulation, both location and consciousness need to exist in order that a generation can function as a vehicle of social change and that active generation units become agencies of change, actively constructing the history of society (Gilleard, 2004; Merico, 2012). Drawing on Wilhelm Dilthey's work, Mannheim also suggests that a thorough understanding of the problems of generations relies on a qualitative appreciation of the personal temporal experience (Costa *et al.*, 2019; Moreno & Urraco, 2018). Each generation, although contemporaneous with other generations, has a distinctive historical consciousness, which leads to a different approach to the same social and cultural phenomena (Pilcher, 1994: 488–489). While older and younger generations may experience the same historical events, the effects of these events will be different due to the different accumulation of experiences along the life course. The gap between young and older generations can thus be explained by the existence of a gap between the ideals they have learned from older generations and the realities they experience.

Mannheim's historical generational theory has been indispensable tools for discussing social change, to overcome the reductionism of cohorts by bringing culture back into the discussion (Aboim & Vasconcelos, 2013), and to emphasise the active human element in the development of history, bringing out the close connection between history and the ways in which people live their lives. This opened opportunities to rethink the relationships that young people have with the wider macro- and micro-processes (France & Roberts, 2015). Notwithstanding this, the conceptual ground for studying generations remains a problematic domain due to the theoretical and methodological difficulties in researching the multiple dimensions of the generational process and the role played by social and cultural factors in the production of social generations (Alwin & McCammon, 2007; Edmunds & Turner, 2005; Eyerman & Turner, 1998; Gilleard, 2004; Pilcher, 1994). Within this context, Aboim and Vasconcelos (2013) underline the need to reach an enlarged conception of social generations as discursive formations, countering Mannheim's reduction of generations to units of individuals operating in the fields of political and ideological struggle. A further criticism is that Mannheim did not provide an answer to the question of how generations act strategically to bring about change (Edmunds & Turner, 2005: 561–562). Instead, it is crucial to understand how sharing a generational experience can produce self-conscious generations acting to trigger change. Moreno and Urraco (2018) believe it is necessary to introduce the component of social class. The two authors recognise the validity of the approach proposed by Karl Mannheim, in its hypotheses of presenting a generation of young people whose characteristics are different from the previous generation as a result of a process of adaptation to the changing conditions of the present. However, they also believe that the degree of exposure to a certain experience depends on the position occupied in the unequal social structure. This position will condition resources, possibilities and ways to respond (more or less successfully) to the various challenges.

Recent theories

Mannheim's authoritative work acted as a basis for the development of further generational theories. A well-known example is the one proposed by North American researchers William Strauss and Neil Howe (1991, 1992). The two authors have popularised the idea that people who grew up and came of age during a particular period in history tend to share a distinct set of values, attitudes, interests and behaviours. Their definition of a generation was inspired by the work of various writers and social thinkers, from ancient writers such as Polybius and Ibn Khaldun to modern social theorists such as the already cited José Ortega y Gasset, Karl Mannheim, John Stuart Mill, Émile Littré, Auguste Comte and François Mentré. Taking Karl Mannheim as a reference point, if Mannheim's theory of generations suggests that generations change in response to major historical events, Strauss and Howe's work favours a cyclical nature of archetypical generations and focuses on the influence of past generations. Moreover, Strauss and Howe (1992) not only looked at how events in history can mould a generation within a specific context, but also investigated the concept that these generations themselves might shape other generations (Knight, 2009). According to Strauss and Howe, each generation has its own biography, shaped by its age location: individuals born at about the same time experience similar epochal events that occur during their life (e.g. Great Depression, World Wars I and II, the Vietnam War, the Energy Crisis). During each stage of the life course, a set of collective behavioural traits and attitudes is produced. The two authors call this a 'peer personality', which they define as a 'generational persona recognised and determined by (1) common age location; (2) common beliefs and behaviour; and (3) perceived membership in a common generation' (Strauss & Howe, 1992: 64). This personality subsequently shapes other generations: specific attitudes affect how children are socialised and later how those children raise their offspring (Coomes & DeBard, 2004; Pennington-Gray *et al.*, 2003). Although beliefs and behaviour are rarely uniform across all members of a generation, those who differ from the peer norm are usually aware of their lack of conformation. Thus, to be a generation, its members must recognise it as distinct from other generations. What leads to this recognition is the interaction the members of a new generation have with members of previous generations and how they experience historical events. This two-part interplay of one generation with another and with important sociohistorical events is defined by Strauss and Howe as a 'generational diagonal'. The generational diagonal acknowledges that generations move through time influencing and being influenced by both important events and other generations (Coomes & DeBard, 2004: 8–9; Howe & Strauss, 2000). Similarly, the importance of chronological interdependencies between generations is highlighted by Joshi *et al.* (2010). The authors propose that each generational identity is associated with a preceding and succeeding generation that are chronologically linked and that this 'unique location of a generation in a chronological order gives each generation access to a set of skills, knowledge, experiences, and resources that can potentially be passed on to or exchanged with the succeeding generation' (Joshi *et al.*, 2010: 393).

Strauss and Howe also characterise historical generations by means of cyclical changes called 'turnings' (Galland, 2009; Sajjadi *et al.*, 2012; Statnickė, 2019). According to Strauss and Howe (1997: 3), Anglo-American society has entered a new turning (a new era) every two decades. At the start of each turning, people change how they feel about themselves, the world and the future. Turnings come in cycles of four, and each cycle spans the length of a long human life, roughly 80–100 years. Together, the four turnings comprise history's seasonal rhythm of growth, maturation, entropy and destruction. The first turning is a 'high', an era of strengthening institutions and weakening individualism, when a new civic order implants. The second turning is an 'awakening', an era of spiritual upheaval, when the social order comes under attack from new values. The third turning is an 'unravelling', an era of strengthening individualism and weakening institutions, when the old civic order decays and a new values regime implants. The fourth turning is a 'crisis', a decisive era of upheaval, characterised by the replacement of the old civic order with a new one. Each generation can be linked to the recurrent sequence of turnings: a 'prophet' generation is born during a 'high' (one key example is the baby boomer generation); a 'nomad' generation is born during an 'awakening' (Generation X); a 'hero' generation is born during an 'unravelling' (millennials); an 'artist' generation is born during a 'crisis' (Silent Generation). According to Strauss and Howe, earlier generations have the greatest influence over new generations. Thus, the generation that emerges after the fourth generation is much more similar to the first one rather than the last generation in terms of values and worldviews. The authors argue that the basic length of both generations and turnings – about 20 years – has remained relatively constant over centuries; however, rapid increases in technology in recent decades are shortening the length of a generation. A thorough examination of the cyclical rhythms of history provides the basis for predicting future turnings. Strauss and Howe's theory has been criticised as a poor use of the concept of generation proposed by Mannheim, an example of overgeneralising and as being unsupported by rigorous evidence. According to Aboim and Vasconcelos (2013: 7), the result of their exercise translates into the production of a plethora of labels (such as Generation X and millennials) lacking specific meanings and often precise time locations.

There is also concern because the different cultures that emerge in each generation makes it difficult to identify one main generational identity. Nothwithstanding this, the universal application of these labels suggests that, as we will shortly see, concepts such as baby boomer, Generation X and millennials are today globally pervasive and help highlight the present and future challenges posed by the generational turnover to tourism (Gardiner *et al.*, 2013). These terms today serve as a shorthand reference to describe the characteristics of different generations (Wyn & Woodman, 2006).

A second example is the work by Ron Eyerman and Brian Turner (1998). The two authors redefined Mannheim's original definition by means of Pierre Bourdieu's (1980) notion of habitus, with the aim of providing a framework for the comparative study of generations. In Bourdieu's (1980: 86) words, habitus refers to 'a subjective but not individual system of internalised structures, schemes of perception, conception, and action common to all members of the same group or class'. As an individual internalises collective cultural practices, these practices become 'embodied' and structure a person's attitudes, expectation and worldviews. The work by Eyerman and Turner (1998) also looks at the interactions between generations and other social categories, including gender and ethnicity, exploring both intra-generational conflict and continuity and considering the circumstances under which generational consciousness may become more salient. This work was further developed by June Edmunds and Bryan Turner (2002, 2005): the emphasis on the struggle for resources distinguishes their theory from the one formulated by Mannheim. Edmunds and Turner (2005: 561–562) suggest that Pierre Bourdieu's (1990, 1993) work on cultural change provides a useful connection – even if in Bourdieu's oeuvre the theme of generations was not systematically discussed (Purhonen, 2016) – to show how intergenerational competitive struggle over various resources, especially in the cultural sphere, produces significant cultural transformation. Generations are seen as becoming politically active and self-conscious when they are able to exploit (political/educational/economic) resources and become culturally, intellectually or politically innovative. Edmunds and Turner (2005) have also suggested that the sociology of generations should develop the concept of 'global generations'. This necessity of a new concept is based on the argument that, with continuous technology developments, the global spread of information and communications technology (ICT) and the pervasive nature of the globalisation process, different cultures are affected by major global events in the same way. Global communications technology has enabled traumatic events to be experienced globally. As stated by the two authors, the late 19th and early 20th century was the era of international generations, united through print media. The mid-20th century saw the emergence of transnational generations, that had access to radio, which speeded up interactions across borders. The latter part of the 20th century is the period of global generations. Global consciousness is supported by the growth of global interdependence, through new communications technology and increasing interactivity, and the 1960s generation was, in this sense, the first global generation. Its global reach was demonstrated by protests and the extent of cross-national activism with a focus on both national and international issues. Due to major developments in global communication, especially in the electronic media, for the first time in history the 1960s generation's very existence was simultaneously diffused to several Western societies, which enabled political and cultural alliances to be forged across borders. This generation provided the initial impulse towards globalisation (Edmunds & Turner, 2005: 564–566).

We finally recall Aart Bontekoning's generation theory. Relying on previous studies – such as those by Ortega y Gasset (1933); Mannheim (1952), Marías (1970), Howe and Strauss (1991) and Becker (1992a, 1992b) – Bontekoning (2011, 2018) established a new theory of generations that tried to link the main perspectives on generations in the social sciences. The author relied on the assumptions made by previous research that generations can be treated as subcultures that have an evolutionary function. The essence of Bontekoning's (2011, 2018) theory is that new generations can be seen as key players in the evolution of social systems. According to the author, generations are shaped by people who feel connected with their peers and are born in a certain time frame: (1) They share life history, circumstances and the impact of historical events, in other words, 'zeitgeist' (the spirit of the time). This creates shared ground for the collective development of a new generation. (2) The most important source is their shared reaction to the actual 'zeitgeist' based on the ability to feel which part of the cultures of the surrounding social systems, such as organisations, needs to be renewed. (3) They develop a shared entelechy, a mix of a shared collective destination and a collective development of mental, emotional and physical attitudes and skills. The common goal is to create change in social systems, such as societies, families and organisations. Lyons and Kuron (2014) assume that, although Aart Bontekoning's generation theory did not receive much attention in the scientific literature, it offers important directions for future research.

Generations and Tourism Preferences

Generational research has investigated the relationship between formative years and significant economic, political and social events that can shape young people's worldviews, values and behaviours. Research has highlighted that major events such as wars, changes in the state of the economy, natural disasters, and political and technological changes can have a substantial impact on individual and collective memories, particularly if encountered during formative years (Becker, 1992a, 1992b; Dencker *et al.*, 2008; Hicks & Hicks, 1999; Mackay, 1997; Meredith & Schewe, 1994; Schewe & Meredith, 2004). Studies express the concept of generational change in terms of unique beliefs, values, attitudes and behaviours being produced by shared histories at the macro level. For example, the individuals who in their formative years fought in World War I shared the experiences of war including death, violence, honour, intense comradeship and the impact of modern technologies. Such experiences profoundly shaped their lives. Although the beliefs and behaviours of a generation are rarely uniform across all members, each generation is expected to display analogous behavioural patterns that are similar among themselves but unlike the previous and subsequent generations (Meredith & Schewe, 1994; Schewe & Noble, 2000). Generational research also established that formative influences that define a generation may have a significant effect on future decision-making.

Recent years have seen an increased interest in generational analysis in the tourism literature (Beldona *et al.*, 2009; Benckendorff *et al.*, 2010; Chiang *et al.*, 2014; Cooper *et al.*, 2019; Gardiner *et al.*, 2013, 2015; Haydam *et al.*, 2017; Huang & Lu, 2017; Li *et al.*, 2013; Pennington-Gray *et al.*, 2002, 2003; Southan, 2017). Studies have considered macro influences in the context of tourism and the consequent effects on tourism expectations and practices. Research has identified that the values and beliefs that are dominant among members of different generations can have a significant impact on tourism demand in terms of motivations, information sources, choice of travel and trip-planning, travel booking, destination perception and choices, travel patterns, preferred activities and destination evaluation criteria (Beldona *et al.*, 2009; Furr *et al.*, 2001; Gardiner *et al.*, 2013, 2014; Glover & Prideaux, 2006; Huang & Petrick, 2010; Monaco, 2018a; Pennington-Gray *et al.*, 2003; Richards & Morrill, 2020). Research has also highlighted significant differences in tourist behaviours between two or more specific generations in different countries (such as Australia, Canada, China and the US), for example between baby boomers and their predecessors, the Silent Generation, or between baby boomers, Generation X and Generation Y (Chiang *et al.*, 2014; Gardiner *et al.*, 2013, 2014; Huang & Petrick, 2010; Lehto *et al.*, 2008; Li *et al.*, 2013; Singer & Prideaux, 2006). The study by Gardiner *et al.* (2014) also indicates that future travel behaviour will differ between the generations and that an understanding of generational shifts in tourist behaviour facilitates the effective prediction and accommodation of future tourism trends.

As recalled above, the well-known study by Strauss and Howe (1992) discovered a pattern in the way different types of generations follow one another in time. In the 20th and 21st centuries, it is possible to recognise several generations with specific values and beliefs that provide the basis for numerous tourism market strategies: the Greatest Generation; the

Silent Generation, the baby boomer generation; Generation X, the millennial generation, Generation Z and Generation Alpha. Each generation displays unique traits and are different in many ways from earlier generations because it has, as explained above, a specific sociohistorical location and a specific identity shaped by cultural experience.

The Greatest Generation commonly refers to those who were born in the 1900s through the 1920s. This generation became the subject of a best-selling book by Tom Brokaw (2004). This is the generation that both came of age during the Great Depression (1929–1939), fought and won World War II and experienced the years of prosperity that followed: rapid social changes, the evolution of mobility and the expansion of mass media. On the one hand, according to Brokaw (2004), members of the Greatest Generation were united not only by a common purpose, but also by common values such as duty, honour, economy, courage, service and love of family and country. On the other hand, rapid industrial and economic growth accelerated consumer demand and introduced significantly new changes in lifestyle and culture. The combination of increasing wealth and technological innovations resulted in the booming popularity of entertainment such as sports, movies and radio programmes. The Greatest Generation members married in record numbers and gave birth to another distinctive generation, the baby boomers.

The Silent Generation follows the Greatest Generation and precedes the baby boomers. This generation is generally defined as people born from 1928 to 1945. The Silent Generation received its name from a 1951 *Time* article that was meant to reflect this generation's withdrawn, cautious and 'silent' characteristics (Hansen & Leuty, 2012; Strauss & Howe, 1991). The Silent Generation (ages over 75) grew up during World War II and the Great Depression, a severe worldwide economic downturn that took place mostly during the 1930s, beginning in the US (the timing of the Great Depression varied across the world; in most countries, it started in 1929 and lasted until the late 1930s). It is a small age cohort because many couples delayed having children during the economic crisis caused by the Depression and also because the upheaval of World War II contributed to the extremely low number of births in this period. The imprint of these momentous events can be seen in the tendency of the Silent Generation to be frugal and cautious or risk-averse, leading to conformity (Javalgi *et al.*, 1992; Pederson, 1992). This generation is defined by its value of social duty as a result of the shared experience of enduring and overcoming those precarious years (Mitchell, 2002). These individuals are described as being very loyal, having a strong faith in institutions and often planning on working for one organisation for a long time (Lancaster & Stillman, 2002). They view work as a duty and an obligation (Kupperschmidt, 2000). Such characteristics mean that this generation is less likely to experiment in their travel behaviour. The motivations for travel among this generation tend to be driven more by planned leisure trips for rest and relaxation, in association with friends and family (Backman *et al.*, 1999). Silent Generation tourists are conservative and are likely to use traditional travel information sources, such as travel agencies or tour suppliers. The study by Pennington-Gray and Lane (2001), which aimed to analyse whether this generation could be segmented into specific types of travellers, profiled the travel preferences of the Silent Generation. The most important preference dimension for members of the Silent Generation was the environmental one. This dimension includes safety, standards of cleanliness and hygiene, weather and easy access to healthcare facilities. 'Education' (which included items such as 'opportunity to increase one's knowledge', 'variety of things to see and do' and 'historical places and buildings') was the next most important factor for members of the Silent Generation. This suggests that the preference for learning while travelling is a large component of the older generations' travel preferences, especially among women. This finding supports previous research which indicates that women tend to place high preference on travelling for education (Pennington-Gray & Kerstetter, 1999). The third factor was the importance of the mobile vacation dimension and the budget vacation. These dimensions were composed of items that focused on price and value of travel: budget accommodation, campgrounds and trailer parks, and inexpensive restaurants (Pennington-Gray & Lane, 2001: 89). Interestingly, these issues are consistent with Strauss and Howe's profile of the Silent Generation. Strauss and Howe (1991) suggest that this generation tends to be frugal and price sensitive because they were children of the depression and grew up during 'sparing' times.

If very few analyses have been conducted which a focus on the Silent Generation, and even fewer studies have looked specifically at preferences for pleasure travel held by its members, much more attention has been given to the baby boomer generation, the result of a strong post-war economy. Baby boomers were born in the years following World War II, between the early 1940s and the early 1960s. This period was marked by a significant and persistent increase in fertility rates in many countries around the world, especially in the West, as the world economy recovered from World War II and the Great Depression (hence the name 'baby boomers'). Baby boomers grew up in an era of prosperity and optimism, bolstered by the sense that they were a special generation, capable of changing the world (Cordeniz, 2002: 239; Yang & Guy, 2006: 270). Baby boomers are often described as a generation that challenged authority, reshaped lifestyles and institutions and broke the mould of the modern life course (Edmunds & Turner, 2005). Boomers also witnessed some of the greatest social changes in history. Between the 1960s and the 1970s, they lived through and actively participated in political and social transformations such as the sexual revolution, the student movement, the hippie culture, the peace movement and the women's liberation movement, and enjoyed prolonged economic growth and optimism. Because of its large size, boomers are seen as a generation that has been forced to be competitive for resources and opportunities (Eisner, 2005; Hansen & Leuty, 2012; Lancaster & Stillman, 2002). As noted by Bristow (2015c), one of the most important features of the baby boomer generation is its significance as a sizeable and powerful consumer group. The post-war boom gave rise to the emergence of the teenager: a new consumer group made up of young people with relatively large amounts of disposable income and a relatively light load of work or family responsibilities (Gillis, 1974). Research has highlighted many differences between baby boomers and their predecessors, the Silent Generation. As regards leisure and travel, existing studies suggest that the attitudes and lifestyles of the former seem to differ substantially from those of the Silent Generation. Given that boomers had the opportunity to become more broad-minded about political, cultural, racial and gender-related issues than any other American generation before them, Lehto *et al.* (2008: 248) suggest that baby boomers seek more physical stimulation/excitement, adventure and quality family time away from home, whereas the Silent Generation travellers are pursuing relatively more static experiences such as gambling, cuisine, and history and culture.

Generation X members were born after the baby boomers, roughly between the mid-1960s and the early-1980s. The label for this generation was popularised by a 1991 book by Douglas Coupland titled *Generation X: Tales for an Accelerated Culture*. Generation X overtook the first name attributed to this generation, the 'Baby Bust', referencing the drop in the birth rates following the baby boom. The 1970s was a period of economic stagnation for much of the Western world, putting an end to the overall post-World War II economic expansion. The firstborn Gen Xers reached school age as the world economy entered a recession, they grew up with a stagnant job market and limited wage mobility and Generation X parents were faced with an energy shortage, high inflation and high unemployment (Muchnick, 1996). While the Silent Generation is seen as loyal and boomers as optimistic, Generation X is described as sceptical (Eisner, 2005; Lancaster & Stillman, 2002), which may be in part a response to the fact that Generation X grew up during a period of economic downsizing and insecurity as to the future (Davis *et al.*, 2006; Gardiner *et al.*, 2014; Herbig *et al.* 1993), simultaneously witnessing critical events such as the Vietnam War and its far-reaching consequences, the oil crisis, the Persian Gulf War, the spread of acquired immunodeficiency syndrome (AIDS) and growing environmental threats (Losyk, 1997).

Generation X was revolutionised by television and the media that provided this generation with more exposure to world events and pop culture than previous generations (Lancaster & Stillman, 2002). Gen Xers pioneered computers at school and witnessed a revolution in computing and related communications technologies. This generation also grew up amid increased maternal participation in the workforce and increasing divorce rates. This is the first generation where the majority of both parents worked outside the home, thereby experiencing reduced adult supervision compared to previous generations. They had a substantially higher probability of witnessing their parents' divorce – due to the introduction of divorce laws, increases in women's employment as well as feminist consciousness-raising – or job loss, due to the economic crisis. As a result of these experiences, members of this generation are believed to be individualistic, lacking in loyalty and less committed to their employing organisation. They are also likely to see work–life balance as extremely important in their personal lives (Beutell & Wittig-Berman, 2008; Eisner, 2005). Individuals of Generation X have thus developed skills of independence, adaptability and resilience (Thiefoldt & Scheef, 2004). The literature describes Generation X consumers as being savvy and cautious (Gardiner *et al.*, 2014; Herbig *et al.*, 1993; Wuest *et al.*, 2008), although they also want a comfortable lifestyle (Schiffman *et al.*, 2008). They are particularly sensitive to overpricing, they are not necessarily brand loyal and they are willing to try new products if their expectations are not met (Yelkur, 2003).

The term millennials is usually applied to individuals who reached adulthood around the turn of the 21st century. Millennials, currently the youngest adult generation, are women and men born between 1981 and 1996 (Bialik & Fry, 2019). In 2020, approximately 1.8 billion people worldwide, or 23% of the global population, could be considered as belonging to the millennial generation (MSCI [2020] – data sourced from the United Nations '2019 Revision of World Population Prospects'). This is the first generation to have grown up amid tensions between globalisation's benefits and drawbacks. On the one hand, the 1980s saw rapid socioeconomic change due to advances in technology (the first IBM PC was introduced in 1981). Internet protocols and technologies were standardised in the late 1980s and early 1990s. The 1990s – a central period for millennials' formative experience – was characterised by the explosive increase in the number of mobile phones and the public availability of the first web servers: the introduction of the internet can be situated in 1995. ICTs played a key role in reducing cultural barriers, supporting the rise of multiculturalism, increasing the density and frequency of international social interaction and the mobility of people, and enhancing the expansion of travel and tourism. On the other hand, in the 1990s, the world economy was hit by a series of recurring economic and political crises with far-reaching consequences brought about by the process of globalisation in the last few decades: increasing precariousness of labour, environmental issues, climate change, financial crises, poverty and inequalities and a growing sense of insecurity over the future. The positive effects of globalisation are not equally distributed between different populations and regions, some of which are less adaptable than others to entering the global market because of different living standards, environment and financial, political and work conditions (European Commission, 2017). Another controversial aspect is the impact of globalisation on self and identity. In the contemporary world, identity construction has become increasingly complex. The construction of individual identities is situated between global and local flows, between the widening horizons of globalisation and the need for local niches for identity construction, between the sense of belonging to the global community and local cultural values, traditions and ideas (Appadurai, 1993, 1996; Bartoletti, 2001; Robertson, 1992). If human beings construct society and cultures in locally situated contexts, they are, at the same time, deeply challenged by discourses, negotiations, conflicts and threats instantaneously transmitted by global networks (Appadurai, 1993, 1996). Globalisation, in other words, has redrawn the confines between time and space (Giddens, 1990), local and global, and thus social relations, gender identities, family models and the living conditions of women and men.

The historical collocation in a period characterised by growing globalisation, a massive increase in the use of the internet and social media and growing global challenges, explains the characteristics of millennials. The millennial generation has been defined as one that is competent, technological, connected and open to change (Benckendorff et al., 2010; Howe & Strauss, 2000; Rainer & Rainer, 2011; Taylor & Keeter, 2010). Millennials are well educated, information technology (IT) literate (they grew up with 2.0 web technologies) and able to multitask. Today's young adults are much better educated than their grandparents. Pew Research Center data show that millennials have higher levels of post-secondary education than earlier generations (Bialik & Fry, 2019). Among American millennials, around 4 in 10 (39%) of those ages 25–37 have a bachelor's degree or higher, compared with just 15% of the Silent Generation, roughly a quarter of baby boomers and about 3 in 10 Gen Xers (29%) when they were the same age. Gains in educational attainment have been especially steep for young women, with younger women obtaining a growing share of university degrees and full-time work compared to older ones. One additional peculiarity of this generation is their racial and ethnic diversity, a result of increasing globalisation and population movements. Millennials have also been raised in more plural family settings, characterised by a growing diversity of arrangements (divorced families, blended families, couples living apart together [LAT], single-parent families and lesbian, gay, bisexual and transgender [LGBT] families) (Parker et al., 2019). With exposure to more cultures, people and family forms, millennials are more tolerant than other generations of a wide range of non-traditional behaviours related to marriage and parenting (Taylor & Keeter, 2010). They are more likely to support same-sex marriage, to identify as LGBT, and are one of the groups most supportive of LGBT rights. However, millennials are also closely interconnected with their families and friends. A 2010 Pew Research Center survey found that two thirds of millennials state that adult children have a responsibility to allow their elderly parents to come live with them (Taylor & Keeter, 2010). Most millennials like working in groups and prefer a sense of unity and collaboration over division and competition.

Another important marker for this generation is that many of them have come of age during a very difficult time in the global economy. The economic and financial global crises had a significant impact on young adults, forcing them to cope with labour shortages, precarious work, low salaries and high unemployment. It also shifted attitudes. As noted (Hout, 2019), millennials might be the first generation to experience as much downward mobility as upward mobility. Millennials also seem to be pessimistic about the global economy and their future and show one of the highest levels of institutional disaffiliation. The 2017 Deloitte Millennial Survey (Deloitte, 2017)[2] shows how millennials are concerned with a large range of global issues and, particularly in mature economies, have a generally pessimistic outlook regarding economic and social progress. Moreover, they are not as confident as older adults in institutions such as government, religion and churches, political parties, the military and marriage. The 2016 Global Viewpoint Millennial Survey (CSIS-IYF, 2017), which surveyed more than 7,600 youth ages 16–24 years old in 30 countries, shows that 67% feel that their government does not care about their wants and needs. As we will see in Chapter 2, because the economic and financial crisis of the last years has deeply affected this generation and its lifestyle, millennials are increasingly price-sensitive and strategic about how to manage their finances (Bernardi & Ruspini, 2018; Veiga *et al.*, 2017). They are looking for more in life than 'just a job': sustainability, personal development, flexible working hours and work–life balance are more important than financial rewards (Buzza, 2017; Cone Communications, 2016; Insead, Head Foundation and Universum, 2014; PwC, 2011, 2015; Telefónica, 2013). In regard to tourism, millennials, more than previous generations, are making travel a priority: use technology and social media to make savvy travel-buying decisions, are increasingly in search of sustainable, experience-based tourism practices, care more about cultural experiences and have a growing interest in solo travel (Barton *et al.*, 2013; Corbisiero & Ruspini, 2018; Sofronov, 2018). A survey conducted by the Boston Consulting Group (Barton *et al.*, 2013) also reveals that American millennials not only value diversity, embrace a global perspective and are open to new experiences, but also seem to be more affected by adverse economic conditions than non-millennials. Most millennials cannot afford extensive leisure travel, and few are fully active business travellers, yet.

Generation Z (also known as the Net Generation or the iGeneration) is a common name for the group of people born roughly from the end of the 1990s through the early 2010s, a span of 15–20 years in the very late 20th and very early 21st centuries. In 2020, Gen Z is comprised of just under 2 billion people globally, around 26% of the world's population (Cushman & Wakefield, 2020; data sourced from the United Nations 'World Population Prospects' 2019). This generation comprises students who are currently at school and university, the children of Generation X. Gen Z members grew up in an era of mobile devices, smartphones and social websites (see Chapter 3 for details). The 21st century has so far been an era of massive technological innovation and Gen Z was the first generation to have ICT readily available at a young age. Technology has given these young people an unprecedented degree of connectivity among themselves and with the rest of the population (Dell Technologies, 2018; Seemiller & Grace, 2016). They are accustomed to multiple information sources and are continuously connected through smartphones, tablets and the internet of things (IoT) (Kardes *et al.*, 2014). Pew Research Center data (Anderson & Jiang, 2018) show that, today, smartphone ownership is nearly universal among teens of different genders, races and ethnicities and socioeconomic backgrounds. Gen Z members are multitaskers and expect to access and evaluate a broad range of information before purchasing (Wood, 2013). They respond more to visual communication such as photos, videos and rich content compared to text. What Gen Zers seem to have in common is an intuitive relationship with social media and digital tools that are continuously evolving. However, they also recognise the need to take a break from the internet's immersive influence and reach a balanced lifestyle while living in the digital age (JWT-J, 2019; Wattpad, 2019). They have been portrayed as a generation whose members have short attention spans (Williams & Page, 2011). As Dimitriou and AbouElgheit (2019) note, this may be a stereotype: it is important to consider that while the constant use of technology has overall shortened our attention span, technology is reducing the time needed to be attentive.

Generation Z is entering adulthood in an era of economic crises that are undermining global stability, experiencing growing fears about emerging social issues such as climate change, sustainability, health and migration crises, and terrorism (Robinson & Schänzel, 2019). The combination of hyper-connection and critical global issues has made them deeply concerned about the future of our planet: they show unprecedented awareness about humankind's impact on the environment and a strong sense of responsibility towards the global community and ecological balance (Wunderman Thompson Commerce, 2019; ETC, 2020). If Gen Zers are frequent travellers – they view travel as a priority – they are at the same time aware and concerned about overtourism and the impact of tourism activities on climate change, and are willing to take a degree of responsibility in mitigating these phenomena by making better booking and travel decisions. Gen Zers have, indeed, been portrayed as socially and environmentally conscious, wanting a mobile-first approach and desiring authentic local experiences (Haddouche & Salomone, 2018; Wee, 2019). Members of Gen Z are also quite young, still developing their own values, habits, beliefs and financial independence (Carty, 2019).

Gen Alpha is the generation succeeding Generation Z (children born after 2012). Influenced by their millennial parents and Gen Z role models, this rising generation is likely to be characterised by strong ethics and values. This is another generation strongly concerned about the environment and open-minded about diversity. A 2019 report from Wunderman Thompson Commerce (2019) reveals that 67% of 6–9 year olds say that saving the planet will be the central mission of their careers in the future, joining the fight that current Gen Zs are leading. This generation already plays a role in the consumer choices of their millennial parents. Gen Alpha tourists may be young, but their ideas and opinions are already influencing family travel decisions. Findings from research from Expedia Group Media Solutions, which explores family travel across nine countries,[3] indicate that, despite their young age, Generation Alpha plays an active role in family travel inspiration and planning. The research report 'Gen Alpha and Family Travel Trends' (Expedia Group Media Solutions, 2019) found that children influence the choice of destination and other components of a family trip such as the activities to do on the trip, trip length, hotel selection and how to get to the destination. The study also revealed that 8 in 10 travellers said that planning a family trip is a combined effort for the entire family, while 60% of those surveyed said that travel ideas come from both children and adults.

Conclusion

The aim of this chapter was to retrace the fascinating history of generational theories and to underline the importance of studying the travel behaviour of generations. Both tourism practitioners and academics have acknowledged the validity of using generational analysis to study generations' travel behaviour (Gardiner *et al.*, 2013; Li *et al.*, 2013). Generational transitions are important moments during which tourism-related experiences change their appearance, taking on completely new configurations. As shown in this chapter, the values and beliefs that are dominant among members of different generations can have a significant impact on tourism needs, values and choices. Millennials and Generation Z travel more than any other generation and young travellers are interested in authenticity, fulfilment and sustainability (Barton *et al.*, 2013).

Still, some theoretical and methodological precautions are necessary. Firstly, it is important to keep in mind that generational theories are useful to define trends, but not to define individuals. As Vincent (2005: 583, 595) notes, the cultural mode of interpretation for social and historical events and changing life chances is subject to change and do not merely stem from an already established generational culture or 'entelechy'. That is, having been a child during World War II does not automatically make a person a conservative-voting nationalist and being born at the end of the 1980s does not necessarily make a person an experiential traveller. As written (Pew Research Center, 2015), the lines that define the generations are useful tools for analysis, but they should be thought of as guidelines, rather than rigid distinctions. However, an appreciation of how the specific generational understanding of major events translates into a set of attitudes helps us understand the salience of these issues.

It is also crucial not to oversimplify: a number of scholars have warned against the attempt to make generalisations about entire generations (e.g. Furstenberg, 2017). Dencker *et al.* (2008) suggest that heterogeneity within generations may be as much as between them. Similarly, the already cited survey conducted by the Boston Consulting Group on US millennials (Barton *et al.*, 2013) showed that millennial travellers are by no means a wholly homogeneous group: a number of distinct segments were identified. For example, the youngest millennials may have more in common with older Gen Zers than they do with the oldest millennials. The study by Skift Research (Carty, 2019) has broken generational cohorts into four smaller groups among which certain characteristics exist: (1) Young Gen Zers (ages 16–18), likely to be in high school and live with their parents, with the possibility of working a part-time job. Most travel is with immediate family (parents and/or siblings) or possibly with friends. (2) Old Gen Zers (ages 19–22), likely to be living away from their parents, pursuing higher education and/or working more often, up to full-time hours. Most travel is with immediate family, friends and possibly with significant others. (3) Young millennials (ages 23–30), likely post-college/university students, working full-time jobs and may be starting families. Travel companions begin to shift towards significant others, spouses and children. (4) Old millennials (ages 31–38): the majority are married, have children and work full-time jobs. Most travel is with their own significant others/spouses and their own children.

Thirdly, we should not forget that the terms 'millennials' and 'Generation Z' themselves are predominantly a 'Western' way of defining generations and that the existing studies tend to consider most millennials/Gen Zers from a limited number of Western countries (particularly the US and Western Europe). According to Akhavan Sarraf (2019), most previous research has relied solely on the Western (mostly American) experience, without considering the specific cultural and historical conditions of other countries and cultures (Lyons & Kuron, 2014). Even global influences are differently manifested in various national contexts as they play out against unique historical and cultural backdrops (Vincent, 2005). It is thus important to pay attention to the local and national peculiarities in the development of generational groups and to support the idea that the characteristics of a generation, rather than being globally identical, are likely to be specific to a national environment. The influence of social media and the spread of the consumer culture could have different meanings within traditional and non-traditional cultures, as well as within occidental and oriental cultures. As suggested by Parry and Urwin (2011: 90), the generational structure within different countries will not follow the Western or Anglo-Saxon model. For example, non-Western research on generational differences conducted in oriental countries such as China, Japan and Taiwan (Hui-Chun & Miller, 2003, 2005; Murphy *et al.*, 2004) has shown that generational characteristics in Eastern countries are not the same as in the West. The study by Huang and Lu (2017), aimed at exploring China's potential outbound market from a generational perspective, has used different generations. The study recruited a sample of 4,047 respondents covering four generations: the Firm Communist Generation (FCG), the Lost Generation (LG), the Reform and Opening-up Generation (ROG) and the Only-child Generation (OG). Another significant challenge lies in the comparison of studies conducted in various countries (Lyons & Kuron, 2014). The availability of global surveys on generations is today, however, growing (e.g. Carty, 2019; CSIS-IYF, 2017; Dell Technologies, 2018; Deloitte, 2017, 2018, 2019, 2020, 2021; Nielsen, 2017; PwC, 2011; Telefónica, 2013; Wyse Travel Confederation, 2018).

Although different scholars have sought to explicate the unique generational configurations of various nations (Deal *et al.*, 2012), it remains the norm to simply adopt the popular US generational categories defined earlier. According to Lyons and Kuron (2014: 142), Mannheim's (1952) influential theory posits that generations take shape within a specific sociohistorical location, making it inappropriate to impose the generational configuration of one society onto another. As noted by Parry and Urwin (2011: 93), the future of generational analysis should consist of not only a longitudinal study by birth cohort, but also a focus on heterogeneity within a generation, taking into account the peculiarities of specific groups, such as women and ethnic minorities, and of local and national cultures.

Notes

(1) Wilhelm Pinder, in the tradition of modern art history, suggested the concept of 'entelechy'. According to him, the entelechy of a generation is the expression of the unity of its 'inner aim', of its inborn way of experiencing life and the world (Mannheim, 1952: 283).
(2) The 2017 Deloitte report is based on the views of almost 8,000 millennials questioned across 30 countries. Participants were born after 1982 and represent a specific group of

this generation: those who have a college or university degree; are employed full-time; and work predominantly in large, private-sector organisations.
(3) Quantitative online survey conducted in the following countries: Australia, Brazil, Canada, China, Germany, Japan, Mexico, the UK and the US. The survey was conducted on a sample of 9,357 people (more than 1,000 in each of the nine countries) between 11 April and 7 May 2019. People included in the sample have children or grandchildren aged nine or younger and had to indicate that they had booked travel online for leisure in the past year.

2 Capturing the Future Traveller

Salvatore Monaco

Introduction

Drawing lines between one generation and another is not an easy task. As explained in Chapter 1, generations must not be identified as labels with which to oversimplify the differences between groups. Rather, they represent a lens through which to understand the social changes that have taken place over the years. The attitudes and preferences of the younger generations are not yet well known. Social change has been rapid in recent decades, generating a series of profound transformations brought about by technological development, globalisation, environmental and health crises (e.g. Adamy, 2020; Gharzai *et al.*, 2020; Oswick *et al.*, 2020).

However, previous research conducted mainly in the field of marketing has highlighted some characteristics and personality traits of the younger generations' members. Their profiles have been shaped by events, albeit in a different way, with an influence over their attitudes, opinions, inclinations and preferences, even in the tourism sector. In other words, these studies can be a useful tool with which to understand what millennials have become, but looking even further, what the main features of members of Generation Z and Alpha will be, too.

The aim of this chapter is to define the profile of millennials, of members of Generation Z and of those younger, starting from relevant studies and reflections that have been conducted both globally and in specific territorial contexts (Corbisiero, 2020; Duffy *et al.*, 2017; Elliott & Reynolds, 2019; McKinsey, 2018). We try to identify the main characteristics of young people, in order to understand some global trends, with a particular focus on tourism behaviours. The literature on the topic has long indicated that tourism is an important element for young people to know the world (Biella & Borzini, 2004; Urry & Sheller, 2004), enrich personal experiences and visions (Gilli, 2015), get in touch with other cultures (Appiah, 2007; Crouch *et al.*, 2005; Rojek & Urry, 1997) and strengthen their identity (Gilli, 2009; Marra & Ruspini, 2010; Nocifora, 2008; Urry, 2003; Wang, 2000).

What kinds of tourist experiences are young people looking for? How are young travellers attracted to a tourist destination? What are the main differences between the various generations? These are just some of the open questions that we try to answer.

Being Young Today: A Social Identikit

As novel parents and teachers, millennials are passing on their ideals and knowledge to the younger generations, even as it is crystal clear that each generation has their different outlook on life, values and habits. Millennials are more family centred than their counterparts of previous generations (Shridhar, 2019). They show intense involvement in the family: they love being with their children and sharing experiences and undertaking activities with them. In most of the young families of economically developed European countries, the male breadwinner–female homemaker traditional family model is now obsolete and gender bias is deeply questioned in many areas of the world (see Chapter 5 for further details). Younger parents are educating their children to not develop gender stereotypes. Very often, men take care of the shopping and cooking; sometimes they attend to the housework and take care of the house and children. Empirical evidence of this attitude is represented by the growing amount of paternity leave which is much more requested by men when they become fathers than their counterparts in the past. Women, on the contrary, are more oriented towards pursuing a career. This is also demonstrated by the fact that there is a constant increase in the average age of first pregnancy worldwide. This is partly because their entrance into the workforce is delayed by their academic plan, and partly because their pursuit of work precedes child-bearing in an increasingly competitive and sometimes precarious job market (e.g. Barroso *et al.*, 2020).

There has also been a change in the way millennials start a family in the Eastern world. In China, for example, millennials were born in a special historical period. With the implementation of the policy of reform and opening up, the year 1980, during which the first cohort was born, marks a transition from the old to the new: old tradition and thoughts were not completely vanishing amidst new ideas and things which were introduced from the Western world. Economically, Chinese millennials have also witnessed economic growth and rapid urbanisation while growing up. With new ideological trends and better living standards, millennials show some differences in terms of outlook on marriage and life as well as consumption. Unlike their predecessors, they pursue freedom in marriage. In the 1960s and 1970s, arranged marriages prevailed in the sense that young people got married not out of romantic love but out of an arbitrary decision made by a third party, often either parents or a matchmaker. From the 1980s onwards, however, the marriage mode started to change gradually as new thoughts were introduced. Young people were no longer 'made to marry', but could marry at their will. Parents' advice was only a 'suggestion', not an 'order'. Also, in terms of 'divorce', according to Li (2019), 1980 also marked the beginning of 'the divorce boom' due to an amendment to the marriage laws in China.

Globally, compared to millennials, members of Generation Z appear more attentive to the environment and social issues. Gen Zers are inheriting from millennials the important awareness to engage in life to achieve important goals, also thanks to the support of technologies, which nowadays have become indispensable (Varkey Foundation, 2019).

It cannot be ignored that members of these generations have witnessed global financial, economic and climate crises. Some studies (e.g. Jaskulsky & Besel, 2013; Yeoman, 2008) have highlighted that especially the youngest generations live with anxiety as a consequence of global warming and climate change. This concern makes young people of the world more attentive and sensitive and encourages them to assume more responsible behaviour. Consequently, even with different intensity depending on the territorial contexts, young people from all over the world, thanks to global real-time communication realised by high-performance technological devices and facilities and high-speed internet connection, have been socialised from an early age to the global difficulties. For example, many demonstrations that have taken place in the past two years testify that members of Generation Z have started to divert their attention towards environmental issues. Millions of young people around the world participated in Greta Thunberg's 'Fridays for Future' events and other similar initiatives, such as 'Run for the Oceans', thanks to which funds were raised for initiatives to protect the oceans and to raise awareness about disposable plastic consumption. According to the study 'Harvard IOP Youth Poll 2019' (IOP, 2019), more than 70% of Generation Z's members think that climate change is a problem; 66% of them consider climate change a crisis that requires urgent countermeasures, not only locally but also internationally.

It cannot be overlooked, moreover, that the health emergency Covid-19 has globally generated a series of economic downturns which have exerted and will exert a strong influence on this generation. From this experience and against this backdrop, the younger part of the population has learned that nothing can be taken for granted, that the world of work is very vulnerable and that an economic crisis can come at any moment. However, this does not mean that Generation Z is distressed or unhappy. By contrast, members of Generation Z are simply more realistic and in some ways appear to be more resilient and determined (Deloitte, 2021). They learned that having a 'plan B' in life is very important so that they can take precautions against unpleasant events and difficulties and cast an anchor windward. This risk awareness pushes them to use new technologies not only as communication tools, but also as resources to increase their skills, broaden their knowledge and raise their consciousness. In this sense, the internet plays a central role in their lives: 41% of the younger generation spend more than three hours of their free time online (Sparks & Honey, 2019).

It is safe to draw the conclusion that there is no clear-cut division between the real world and the virtual world for Gen Zers. This generation is so hyper-connected that they move with elasticity from one dimension to another: real and digital are two worlds that complement and interpenetrate each other, mutually conditioning. In this scenario, a central role is played by social networks, which have taken on new forms and meanings. If for millennials, social networks are tools for sharing posts, pictures and news with their contacts, for members of Generation Z, they represent a source of inspiration, a space in which to interact with other users, evaluate different options, compare prices and understand their peers preferences. Highly sought-after social network platforms include Instagram, TikTok, Reel, Triller and live streaming platforms such as Twitch, in which users communicate mainly through images and videos.

Main Consumer Choices of New Generations

The characteristics of the younger generations described in the previous pages are also reflected in their preferences and choices of consumption in a number of sectors, including technology, food and tourism.

As we will see in the following chapters, millennials by 2025 are estimated to constitute 75% of the global workforce, and today prefer consumption to experience in the company of their families instead of consumption to own (Euromonitor International, 2019). Although it has long been theorised that millennials were self-centred (Gillespie, 2014; Koczanski & Rosen, 2019; Konrath et al., 2011; Twenge, 2013), the happiness of children is an important consideration in the consumption of contemporary young parents.

Hence, the philosophy of the sharing economy, which has characterised the life of millennials in recent years (and which we will discuss in depth later), is beginning to extend to family life. Millennials no longer buy cars and show no interest in owning a house or an apartment, instead they are starting to exchange, borrow or rent articles of daily use for their family, such as car seats, clothes, toys and accessories. This outlook on life particularly concerns the Western world. On the contrary, for millennials of some emerging economies, such as India, the possession of cars and apartments still represents a status of success in life. According to the survey of Airbnb China and CBN Data, 57% of the respondents consider that 'travelling around the world' is the top priority in their life, in contrast to 49% of those who wish to buy an apartment. Different habits are also registering towards some consumer goods.

In China, however, it is only in recent years that the concept of sharing has been widely spread. In June 2014, the app Xianyu (Idle Fish) was officially available on the iTunes Store. Attached to Taobao, an e-commerce giant in China, it is an online C2C platform where people can buy new or second-hand articles from others or sell their own 'idle items' with just an easy click on their Taobao app. An individual seller or buyer can enjoy the highest exposure from the millions of users on Taobao and the most efficient logistic service from an almost autonomous service (make an appointment on the app and the courier comes within a few hours, all payment and tracking number filling are done automatically). In 2016, Xianyu also included many other 'sharing' functions: users can borrow second-hand books, children's toys, clothing and digital products after they have been renovated and disinfected at the centre of Xianyu. Today, Xianyu has become a multifunctional platform that includes online auctions, live-streaming, house rentals, a social networking community and so on, setting an example of the trend of a sharing economy in China. Attracting almost 200 million users across China, the platform also encompassed a bigger role in the Covid-19 period. Many millennials lost their job due to the pandemic and they had no choice but to make a living on their own. They turned to Xianyu to sell their idle possessions for money or to promote their services such as online tutoring to help them get through the trying circumstances.

CB Isights, an American data analysis research institute with a focus on industries and private companies, recently published an article with the findings of research based on the US market (CB Isights, 2019) which highlight that some industries seem to be more affected than others by the new preferences of millennials and future generations. Especially in America, casual dining or the traditional restaurant have now become one of the favourite places of millennials. Among them are restaurant chains such as 'Ruby Tuesday', 'Olive Garden' and 'Applebee's', which received highly favourable comments for their simplicity and leisure in America in the 1980s and 1990s, where people could enjoy a simple lunch or dinner without spending too much. Unlike their parents and Generation X who preceded them, millennials are more attentive to the quality of the food for themselves and their children: they prefer spending more money on food and be sure of eating well. When they can afford it, they frequent luxury restaurants. Alternatively, millennials prefer to eat at fast-food restaurants that offer quick lunches and dinners, as long as there are healthy options on the menu (e.g. many of them say no to industrial cheese and prefer craft beer to packaged beer).

Another of the main characteristics of the new generations, from millennials onwards, is the need to take action in the form of a dietary shift towards more plant-based diets.

Four of the top five global vegetarian markets are in Asia (with over 500 million consumers across India, Indonesia, China and Pakistan), since vegetarianism has been established in Eastern cultures for centuries (Euromonitor International, 2020). In general, across the Asia-Pacific region, due to religion, altruism, and social and cultural precepts, the size of the vegetarian and vegan population, and the corresponding products from food manufacturers, are definitely on the rise.

The development to follow plant-based diets or any form of vegetarian diet is spreading around the world, and members of the younger generations are the main protagonists of this change in eating habits. In fact, recent studies (Berkhout *et al.*, 2018; Ginsberg, 2017) underlined that the trends of flexitarianism, reducetarianism and part-time vegetarianism are growing in Europe. Similarly, a recent statistical study from America (Blázquez, 2021) stated that, while only 2.5% of Americans over the age of 50 consider themselves vegetarians, 7.5% of millennials and Generation Z have given up meat. The same is true of veganism, where younger generations have taken up this kind of diet at almost twice the rate of older Americans.

The young movement towards vegetarianism and veganism is based on the idea that this food transition will not only benefit people's well-being, but will also prevent the exploitation of natural resources throughout the world.

In addition, the way of shopping has profoundly changed thanks to the popularisation of e-commerce. Thus, as the internet plays an increasingly central role in the lives of young people, their purchasing behaviour is gradually changing. More specifically, their purchasing behaviour is profoundly influenced by the speed of transactions. The most captivating purchases are those that can be made in a very short time. Furthermore, they make great use of mobile devices for shopping. In the past, millennials shopped in offline physical stores and only felt assured when they saw and tried the products themselves. However, with the emergence of specific apps (such as eBay, Privalia, Taobao, JD and Pinduoduo) their purchasing behaviour is gradually shifting from offline to online channels. Young people are fond of buying almost anything on the e-commerce platform, ranging from daily necessities, luxury goods and designer products to automobiles and furniture.

The forced termination of many social and economic activities during the lockdown caused by the epidemiological emergency of Covid-19 has given an important impetus to online sales, which are already widespread among young people (Brosdahl & Carpenter, 2011; Duffett, 2015; Elwalda *et al.*, 2016). It cannot be overlooked that globally Amazon, the largest e-commerce giant, in the midst of the health emergency, received so high a boom in order requests that it was forced to prevent customers from buying unnecessary items to allow for the normal shipment of basic necessities.

In addition, it is safe to say that in the pandemic period, many physical stores survived thanks to the popularity of e-commerce platforms and live-streaming apps (such as TikTok and Douyin). Most traditional offline stores began to transfer their transactions from offline to online and 'hawk' not on the street but in live studios, attracting millions of viewers. According to iiMedia Research, as of 2019, the total users of live-streaming reached 504 million, of which millennials occupy the biggest part. The digitalisation of traditional industry also attracts celebrities and luxury brands. Furthermore, the precautionary measures taken before entering a store and the fear of contagion continues to act as a stumbling block for many consumers, and so they will continue to choose e-commerce as their main way of shopping even after the forced quarantine is ended (Global Web Index, 2020). In addition to home delivery, there is also the model of 'click and collect', where customers can choose to pick up their ordered articles from the nearest physical store after they pay for them online, which has become one of the most widespread practices especially favoured by the youngest generations globally (Osservatorio e-commerce B2C, 2020).

The propensity for online consumption has also been inherited by Generation Z members, accounting for a population of around 2 billion worldwide; by 2025, however, they will make up over 30% of the global workforce (Martínez-López & D'Alessandro, 2020; Priporas *et al.*, 2017). For Generation Z, accustomed to multitasking and the simultaneous use of different devices (such as a smartphone, tablet and laptop) and also online purchasing, access to e-commerce sites is the norm. They also like to frequent shopping malls and physical stores, as long as these places are able to combine physical and virtual offers through apps, 3D multisensory experiences and new possibilities offered by technologies (Lyons *et al.*, 2017).

Recent studies (e.g. OC&C, 2019) have described the member of Generation Z as a powerful driver in the family's purchasing decisions. They are attentive, wary and picky consumers, so much so that they have also been called 'ExperTeens' (Morace, 2016) because they have grown up with the rise of digital devices and are more pragmatic but less self-indulgent than the elder generations. At least in the West, they especially appreciate companies that take political positions and share their social and environmental causes. However, this does not mean that traditional products or brands are abandoned by Generation Z. Instead, they are selected, studied and purchased if they prove to accord with the personalities of consumers.

Things are a little different in the East: in these territories young people seem more carefree and inclined to buy without having too many concerns. This attitude explains why the biggest international brands are adopting diversified strategies to attract the youngest consumers of the Eastern market. The big giants, in particular, are carrying out more aggressive and decisive marketing actions within the Asian markets compared to those carried out in America and Europe. The strong consuming potential has also attracted the attention of luxury brands.

Not only official physical stores and e-flagship stores on e-commerce marketplaces, but luxury brands have also opened their official account on live-streaming platforms. In 2018, Dior was the first brand to open its official account on Douyin, later followed by Gucci and Louis Vuitton. In August 2020, Louis Vuitton for the first time launched its SS2021 Men runway in Shanghai China, and at the same time, went live on the official account of Douyin, attracting millions of viewers across China and the world at large. The brands consider the platform as a channel that helps them establish the strongest emotional bond between traditional luxury brands and young customers who are the active users of these platforms.

The Role of Social Media and Influencers

Technologies, in particular the communication ones, are one of the main elements modelling the different generations. In the Western world, baby boomers grew up in the age when the popularity of television had expanded, thus changing their lifestyles by making it possible to know the world from their home in real time. Generation X grew up as the computer revolution was in full swing and millennials came of age during the internet explosion. They distinguish themselves from previous generations for being part of the first global generation, characterised by unique habits and ways of thinking and for being very familiar with the means of communication and digital technologies. After them came Generation Z, which has grown into an even more globalised group, in which the internet has always existed in a widespread way. They are also termed as the generation of iPhones, high-speed connections and WiFi networks. While previous generations have had to discover and learn the potential of these technologies, from Generation Z onwards these tools are taken for granted. There is no image of the world in their mind that is not hyper-connected and supported by digital technologies. In light of these considerations, it is clear that it is not possible to analyse the behaviours of the various generations without taking into account the influence of modern technologies, as well as the consequent possibilities brought by technological devices in the passage time, such as growing opportunities and prospects.

Today, we are in an 'always active' ever expanding technological environment, in which social networks play a central role. They are the main cyberspace where meetings, exchanges, conversations and shares take place (Boyd & Ellison, 2007; Han *et al.*, 2018). In the e-commerce era, social networks are also starting to be used for shopping: Facebook, YouTube and recently Instagram offer their users the opportunity to purchase products directly from their platforms. In other words, in contemporary society social networks represent real communities that are blurring the distinction between online and offline worlds (Duffett, 2015).

According to data from the periodic survey of the Global Web Index (2019) aimed at analysing the use of social media in the world, internet users spend 20 minutes and more on social media per day than on television. The analysis was conducted on 350,000 subjects worldwide between the ages of 16 and 64. For Generation Z, the difference increases to one and a half hours. In general, Instagram is certainly the most appreciated social network app, especially since the feature of Instagram Stories was rolled out, which offers users the opportunity to communicate not only through images, but also through videos. The second most used social network software globally is Facebook, followed by YouTube. Twitter and LinkedIn also have a certain group of followers, but are mainly frequented by the elder generations to keep up with current affairs and politics, or hunt for jobs. According to the report, 54% of users (excluding China due to its restrictions on the web and social media) have watched at least one video on Facebook, Twitter, Snapchat or Instagram during the 12 months prior to the survey. Social networks have shaped the daily reality of younger people. There are a lot of varied meanings and unconscious strategies behind interactions on these social networks that seem to be apparently simple gestures. Nothing is random behind a like or an emoticon, but the underlying meaning is that they are a way of seeking approval, that they are a strategy to conquer a person or that they are traces left as a sign of friendship (e.g. Cantelmi, 2013; Giorgetti Fumel, 2010; Siegel, 2011). Even not putting a like represents a standpoint: it can be interpreted as disapproval or resentment. Regardless of the social network that young people use, what they seek is something that arouses amazement and that triggers an echo in them (Bolton *et al.*, 2013).

Instagram and TikTok are used by young people above all to observe what others do (friends and others), to maintain friendly relationships with people (acquaintances and strangers), to get current news from the world and to follow the celebrities and influencers whom they are crazy about. If for millennials social networks are also a platform in which to tell others about their experiences, through the sharing of posts, images and videos, members of Generation Z seem more interested in enjoying the contents that are published by others, than in making their own. This does not mean that they do not publish original content, but they do in a more thoughtful way. To further study this aspect, we can refer to other interesting data from the Global Web Index. The study underlines that, in general, members of Generation Z use social networks not only to keep in touch with friends and acquaintances, but also to follow so-called influencers. This term is used to refer to people who become famous online and are followed by young people because they are outside the traditional show business (Liu, 2019; Patterson *et al.*, 2013). They can be YouTubers, TikTokers or have a successful Instagram profile. For this reason, their experiences and advice, including about purchasing, are considered more authentic and less partisan than the messages conveyed by a celebrity through the traditional means of communication. Successful influencers are mainly those able to anticipate trends, arousing the curiosity of followers through a defined style, which can be constantly inspired.

The central role played by influencers today represents an interesting new element which allows to further understand the generational shift. Indeed, the other generations were usually inspired by characters from the traditional world of entertainment (movie stars, singers and television characters) with an age older than their own and therefore with more experience (Lewis, 1992). Today, members of the younger generations rely above all on influencers who are of the same age through comparison with them and emulation of their behaviours.

A similar picture can also be found in China. From QQ (an online instant messaging software) to WeChat (an instant messaging app), Chinese millennials have never diverted their attention from social network platforms. From the 2000s onwards, millennials started to chat with each other on QQ and write articles (a form of blog) in their Qzone on their PC. This was considered as the predominant manner of online socialisation in the first decade of the 21st century. After the company of QQ launched another similar product for mobile terminals, WeChat quickly grabbed a big portion of the market share from the year 2011. Initially, WeChat was only used for chatting between friends and sharing everyday life on Moments (an app similar to Instagram where people can post photos or short videos). Putting a like on the photos or videos shared on the individual feed of Moments is seen more as an act of social courtesy. WeChat also allows its user to make comments on the photos and videos and thus the comments make room for people to interact with each other. In the following years, use of WeChat was also extended to the working environment where colleagues establish a chat group for a specific purpose, for example, discussion or follow-up of an important project. Also, the appearance of public profile and the payment function greatly strengthen the user frequency of WeChat. It is no longer merely a social network, but a channel in which brands attract the attention of customers. Fashion brands and luxury brands successively opened up their public profiles on WeChat, publishing articles or carrying out online activities to strengthen their bonds with customers. Meanwhile, another social network media most followed by Chinese millennials is Weibo (that means Microblogging). As its name suggests, Weibo allows people to write a small paragraph or a micro-blog of no more than 140 Chinese characters. Netizens voice their opinions or share their everyday life by writing or posting photos and videos. Similar to WeChat, the comment area is also used for interaction between friends and families. With its popularity in recent years, brands, celebrities, governmental offices, schools and universities and all other entities of all sectors have opened up their official Weibo accounts where users or fans can interact with celebrities and make their voices heard.

Tourist Choices and Instagrammability

Two important findings emerge from the studies carried out on the younger generations: firstly, on a global level, their interest is directed towards enriching experiences from a human point of view and in line with ethical values and expectations; second, new technologies in general, and content online or in social networks in particular, are now more than ever considered to be modern guides, which provide information about the world, and sometimes also determine choices. These new trends are also reflected in some of the main tourism decisions (e.g. Wu *et al.*, 2008; Yeoman, 2008).

According to research presented by Expedia during Explore18 (Expedia, 2018), the annual convention held in Las Vegas, the holiday preferences of millennials and Generation Z members are not so different in many ways. Both generations are constantly connected, with over five hours spent on the smartphone per day on vacation and very extensive use of social networks. A habit they repeat especially during their travels. This trend was also confirmed by the Europe Online 2020 study (PhoCusWright, 2020), which revealed that in Europe around 20% of young travellers use applications at every stage of travel planning, while over 60% of them have at least one travel application on their phone. The information present on social networks (in particular Facebook for millennials and Instagram for Gen Zers) is the main source for choosing a destination. For 30% of young people, this kind of choice is also conditioned by the votes and the feedback that the various locations receive in online reviews written by others who have experienced the destination before, for example, Yelp. Despite the common ground between members of the two generations, there are differences in behaviour between the two groups.

Millennials are risk-aware. According to the study, they seek travel experiences that are verified through their network or trusted sources. In other words, especially in the tourism sector, millennials prefer not to be the first to try something; they want to know before departure what awaits them once they arrive at their destination. In this respect, members of Generation Z appear more adventurous and resourceful, which is also confirmed by research conducted by Booking.com on over 22,000 participants in 29 countries (Booking, 2019b). The study revealed that travelling is a top priority for Generation Z and 65% of them have already decided to spend part of their savings on tourism for the next five years. For 60% of the young people interviewed, having tourist experiences is even more important than saving money for the purchase of their first home. Generation Z travellers are more persuaded by the photos and images of the places they see than the reviews and opinions of people who have already had the same experience. Some 54% of travellers from Generation Z like to search on the various social networks that contain posts and photos dedicated to travel and 40% of them often go to look at their feed to find inspiration on where to go and what to do on holiday, thanks to hashtags like #travel, #inspiration and #tourism. Women are more inclined than men to plan their future trips in this way, especially as they scroll through Instagram posts, beating 25% of the global average and 30% recorded among millennials. When they have to decide where to go, 45% of Generation Z members say they are convinced by influencers and travel bloggers, with 35% of young people ready to trust their recommendations and advice. However, they love horizontal communication, so even when they admire someone, as in the case of influencers, they expect to communicate with them as equals, as if they were talking with a friend. But inspiration not only comes from the smartphone: 35% of the participants in the study would like to visit destinations seen on a TV series or the movies. And 33% continue to seek advice from friends and relatives, or travel agencies.

By emulating the management of their favourite influencers' profiles, the smartest Generation Z members enrich their personal feed with many well-made photos: 43% of Gen Z confirm that they choose destinations for taking spectacular photos, and 42% always upload some pictures on social media, against 35% of travellers belonging to other age groups. These results are in line with other studies that have focused on a very recent, but already widespread, phenomenon that has been defined as 'Instagrammability'. The word Instagram has recently been extended to be used as a verb among the terms from the British Encyclopaedia (2019) and its derivative adjective 'Instagrammable' has become a label to define what deserves to be published and shared on social networks. To be more specific, an 'Instagrammable location' deserves to be shared on Instagram because it is able to capture the attention of the online community and to receive popular appreciation in terms of likes, comments and re-posts.

Schofields Insurance (2017) has recently conducted a survey on over 1,000 UK people aged between 18 and 33 on this topic. The findings have revealed that the Instagrammability of a place is the biggest motivator for people under 30 when they are preparing their holiday plans.

Social networks exert a growing influence over the choices of younger travellers, reshaping the traditional communication flows of the tourism industry. In other words, Instagram is definitely having an impact on travel destinations. Today, many destinations, once considered very distant or attractive only to adventurers, now catch the eye of many more tourists, some of whom travel for the purpose of taking photos 'worthy of being posted on Instagram'. The Mobile Travel Tracker by Hotels.com (2017a) highlighted that when young people go on vacation they are very attentive to the 'social' aspect of the trip. Selfies, posts and status are updated above all to invite envy from their contacts, triggering a sort of implicit competition between people who are able to capture and share the most beautiful, exotic or suggestive image. This kind of behaviour is defined as 'travel bragging', which refers to the act of boasting about travel experiences. In this sense, likes and comments are considered increasingly important parameters with which to evaluate the success of a holiday and, in some cases, also to try to establish themselves as influencers on the web. The study of Hotels.com reveals that 11% of younger tourists almost exclusively select structures with panoramic views, rooms with particular furnishings, innovative hotels from an architectural point of view, mainly with the aim of creating content to put on the Net in the form of photos, videos or stories that can attract the attention of their followers. In China, most millennials rely on online platforms of tourist experience sharing, such as Mafengwo. They check the recommended trip itinerary and the tourists' reviews before embarking on their journey. Also the photos and short videos shared on the platform work as good resources for them. Then, if they settle on a destination, they can quickly find the channel for reserving a hotel and buying a train or airplane ticket on the platform. In addition, their tourist choice is also determined by internet celebrities and pop stars. Photos and articles posted on WeChat attract the attention of millennials who cannot wait to leave for the destination recommended. In China, where the use of Instagram is prohibited, the term 'Instagrammability' may well be changed to 'wechatability', as youngsters mostly share their photos of beautiful scenery during trips on the Moments space of WeChat and friends or colleagues would put a 'like' as a sign of 'appreciation'.

In Search of Sustainability

It is important to emphasise that social networks are not everything for new generations. In fact, millennials and Gen Zers are both aware of the impact that travel can have on the environment. For this reason, young people are willing to avoid destinations usually crowded by mass tourism, showing a sensitivity towards environmental and territorial sustainability issues. In this regard, the data released by the platform Pinterest are interesting (Deep Focus, 2018). The social network has collected and disseminated a series of information on the research done by its users regarding travel in recent years. The analysis shows that people under the age of 38 have carried out a lot of research on images and ideas related to sustainability and the environment in the world in recent years. Globally, for example, searches for the terms 'sustainable life for beginners' grew by 265% in 2018 compared to 2017. More specifically, in regard to the tourism sector, searches for 'zero-impact travel' increased by 74% from 2017 to 2018. Younger Pinners try to travel lighter, reduce their carbon emission transportation and produce less waste, especially when they are far away from home.

Recently, in the psychological field the term 'eco-anxiety' was introduced (GBD, 2017) to refer to a lasting fear of environmental damage, which affects above all the younger generations. It should not be considered a pathology, but a form of anxiety, which, however, is still without a definition and absent from the psychological manuals. Thus, the rapid and irrevocable impacts of climate change, as well as a lack of attention to the protection of the environment and nature, constitute, without a doubt, a source of stress for many youngsters. Psychotherapeutic studies concerning precisely this issue have been carried out both in the US and in some European countries.

Therefore, it is not a coincidence that an Italian survey (SWG, 2019) demonstrates that climate change is also the top concern of young residents of Italy; 64% of the Generation Z youth sample indicated the climate at the head of the 'concerns that most worry'. To feel less guilty or powerless for the lack of control over nature, many young people make more responsible tourist choices, trying to combat eco-anxiety by exhibiting respectful, sustainable and eco-friendly behaviours. Another research carried out by Booking (2020), based on an analysis of over 150,000 worldwide destinations sought after by its users to find the ideal accommodation, has highlighted that the dream of over 28,000 travellers in the world is the right combination of the tourist experience and contact with nature. About half of all global travellers of any age (51%) choose to reduce their CO_2 emissions by limiting their travel distances. This percentage increases for members of Generation Z once they reach their travel destination: 63% of them said they chose more eco-friendly means of transport, such as a hybrid car or an electric rental car, and public transportation or even prefer bike riding. Consistent with these intentions, 49% of the profiles in discussion evaluated the possibility of visiting natural wonders. At the top of the list there is, for example, the river of five colours in Colombia. This masterpiece of nature, also called the 'rainbow river' or Caño Cristales, is located in the Sierra de La Macarena National Park, which is a small, unspoiled rural reserve. Some 52% of Gen Zers would like to include a trekking experience within their holiday. Of these, 63% have oriented their research towards lesser-known destinations to limit the probable environmental impact brought about by their travel. About half (44%) of the Generation Z travellers surveyed said they would like to combine the travel experience with the possibility of exerting a positive impact on the local community of the destination by volunteering. This particular category of tourists wants to be in contact with local populations, to see how they survive, and in what conditions they work or spend their days. From this perspective, tourism almost takes the form of an activity that travellers practice to look for peace and a state of grace (Heelas & Woodhead, 2005) because they are aware of their own privileges. Therefore, in this case the 'tourist gaze' is oriented by the philanthropic will to show and understand difference. At other times, tourists also engage in socialisation. According to the data collected by hotels.com and hostel world (Hostels.com, 2017b), two of the largest search engines in the tourism sector, among the most sought-after destinations for the discovery of natural beauty, places that are recognised as the heritage of humanity and places of great historical and artistic value are Iceland for a swim in the hot spring of the Blue Lagoon; China for climbing the Great Wall, built on mountains; Egypt for visiting the pyramids; Australia for seeing sunrise from Uluru, snorkelling along the coral reef and admiring the Grand Canyon by helicopter.

Some studies (e.g. Artal-Tur & Kozak, 2019; Cavagnaro *et al.*, 2018; Glover, 2009) also highlighted 'physically demanding' tourist experiences among the new youth trends. With the increasing popularity, the 'physical demands' include cycling in rural areas and various physical activities in mountain regions. For this reason, among the favourite destinations for young people there are also many Chinese metropolises, where a combination of the tourist experience and physical activity is possible. Especially in some areas of ethnic minorities, tourists can feel the local culture by attending the characteristic activities. For example, in Hangzhou, a city famous for its production of the best tea in China, Airbnb has promoted an experience project. Tourists can follow a team grower and learn from him or her the process of plucking the tea and producing the tea as well as traditional tea etiquette. While admiring the breathtaking scenery of the tea garden by the West Lake, tourists also immerse themselves and feel and learn the unique local tea culture.

In line with the exhibition of more sustainable behaviours, it can argued that a real food revolution is underway. Young travellers today are very attentive to the preference for zero-kilometre products. This term refers to the usage of local foods that have not travelled far after production. This principle begins with people's dedication to local and regional specialties. Consequently, while travelling young travellers tend not to look for their own culinary tradition, but choose to try local cuisine. Recent years have seen a sharp increase in the consumption of bio-vegan food during tourism trips: younger generations pay close attention to a quality and nutritious diet. The commitment to eating local specialties also safeguards the conservation of rare food species, since zero-kilometre products keep unique species alive and promote eco-diversity. This approach towards food promotes not only territorial independence, identity and tradition, but also the preservation of the environment and the land where the food comes from.

In conclusion, it is safe to argue that new generations seem to be more socially and ecologically responsible than previous generations. 'Green travel' will occupy an increasingly central place in the tourism market globally, especially since the health emergency brought about by Covid-19 has somehow relaunched tourism in the open air, which can favour the maintenance of physical distances among people (e.g. Corbisiero & La Rocca, 2020; Gössling *et al.*, 2020; Corbisiero & Monaco, 2021). In general, it goes without saying that the younger generations aspire to be more environmentally conscious travellers by making decisions that help protect our planet. This attitude will be increasingly reflected in travel organisations: from the destination to be visited to the means of transport to be taken, from the activities to be carried out on site to the food to eat. It is plausible that hybrid or electric rental cars, public transport and even walking and cycling tours as well as eco-friendly restaurants and accommodation facilities will experience great momentum.

Conclusion

Talking about tourism today means referring to a practice completely different from what it once was a few years ago. If, in the past, tourism marketing was based on the traditional marketing model of 4P (product, price, place, promotion) (e.g. Hayward, 2002), this type of approach today seems to be unable to capture the attention of travellers of the new generations. Thus, for young travellers, the emerging orientation is to seek experiences, emotions and memories in the travel experience in line with their values and principles. These are intangible components, which, in importance, exceed the 'physical' ones of the tourism product. In other words, younger travellers expect a dynamic, modern, well-designed tourism product equipped with high technology. It is therefore inconceivable, for example, that Wi-Fi is not available or that this technology is not free of charge. The possibility of obtaining online information on how to find and learn about the activities offered in the area, for example, check in online, request a check-out delay, book room service, pay the bill online and book a transfer, certainly represents an added value. Travellers of the future seem to be less loyal to the most famous hotel chains and brands, but rather, they are willing to choose the destination and the structure in which to stay considering also the ethical principles in which they believe.

Members of the new generations do not want to maintain 'common' contacts with destinations and structures, but prefer personalised communications and targeted information. As mentioned earlier, the internet in general and social networks in particular are currently the main channels for collecting information. Therefore, a trend that could develop is the application of direct contact channels between young travellers and accommodation or catering facilities. They could take the form of closed online groups to create a relationship and, in a sense, spread the idea of community. In this way, young travellers would have the opportunity to express their preferences, opinions and perplexities and the structures could prepare in advance to welcome travellers from all over the world.

Nowadays, the most innovative companies and organisations are starting to communicate with their customers through digital channels.

These communication channels not only often reduce the time to get a response to the customers, but they also create a more direct and instant relationship between people. Customers can submit their question and be notified when they receive the answer. The favourite channels for young people to communicate with companies is WhatsApp. It is preferred because it is more closed and secure than other apps since all the exchanged messages are end-to-end encrypted and also because there is no connection to a public social profile, but only to the private phone number. While it has not surpassed WhatsApp in number of users, Messenger is more geographically distributed. Furthermore, its use is facilitated by the possibility of connecting it directly (and freely) to the brand page. From this point of view, Messenger is more open than WhatsApp, because its use does not require the possession of a phone number. These features make it easier to use for customer service. The natural advantage of Messenger is that it is connected to the largest social network in the world: Facebook. The first alternative to Messenger and WhatsApp is Telegram. Even if this app is not as globally widespread, it is highly appreciated by young people above all because it protects the privacy and security of its users. Telegram has the option to create secret chats: self-destructing conversations, even if they are not end-to-end encrypted by default like WhatsApp. In addition, Telegram has a rich chat-bot ecosystem, which fits neatly into customer service options. Telegram has not created corporate accounts yet, but businesses can create channels or groups, which can be used for marketing and community purposes.

Members of Generation Z enjoy great digital literacy and for this reason they are sometimes wary of the information they find online, fearing that it could be fake news (e.g. Zimdars & McLeod, 2020) and they often fear that the contents produced by tour operators are biased or false (e.g. Benckendorff *et al.*, 2019). For this reason, they often turn to online tourism communities where they come into contact with their peers and exchange opinions and reviews. In these communities, what I have defined elsewhere as 'tourist communitycation' takes shape: it is a neologism created by the fusion of the words community and communication (Monaco, 2019b). In particular, the tourist communitycation takes place through platforms for sharing reviews, opinions and tourist information among potential travellers, local populations and tourists who interact and support each other. The online comparison, especially with foreign people, is increasingly sought after as young people assume that the opinions exchanged are selfless and thus more honest and reliable (Piotrowski & Valkenburg, 2017). For tour operators, a useful strategy of adapting itself to the new communication flow though currently only in part implemented, is to create direct contact with travellers (Sfodera, 2012). Most likely tourists will triangulate the news, looking for confirmations among different online sources and what other members of their communities say. As a result, structures and destinations that intend to recommend themselves to the younger public should mainly emphasise the verification of news and the combating of fake news. Creating a welcoming, engaging, but also reliable communication channel is an indispensable condition for any tour operator who wants to address the needs of the younger generations.

Therefore, looking to the future, it is very probable that the next generations will centre on the importance of technology and environmental issues in their socialisation. After all, we can already see today how the Alpha Generation (McCrindle & Wolfinger, 2009), which includes people born after 2010, is exposed to an unthinkable quantity and variety of technology used since birth for learning, interaction and play. For the Alpha Generation, mobile devices have always been within their reach. According to the most recent Ofcom (2016) report on the use of the media by children, the number of children aged 15 who own a smartphone is on the rise: 41% of children have a smartphone and 44% have a tablet. If we talk about use but not possession, the number also grows: most children aged 3 or 4 (55%) use a tablet. In addition, the internet of things has now fully entered homes with devices such as Amazon Echo and Google Home, with which all family members who own them are now familiar. They are used to turn on lights, listen to music, put a movie on TV, surf the net, etc. A team of Mit Media Lab researchers recently conducted a pilot study to explore how children interact with artificial intelligence devices (Meltz, 2017). The researchers observed a group of children between the ages of 3 and 10 interacting with Google Home, Amazon Echo Dot and other similar technologies. Artificial intelligence agents are considered friendly and reliable for children. Some of the older children have also claimed that these devices are smarter than they are, with particular reference to Alexa.

As will be explored in Chapter 3, interaction with artificial intelligence is becoming increasingly popular globally, making it possible to overcome the screen–keyboard relationship. Therefore, it is easy to speculate that these technologies will soon have to be fully incorporated into the tourism offer all over the world. In the future, connected devices will become the norm and this will ensure that children expect them to be always present; even when they are away from home, the devices would also respond to their voice commands. The propensity towards technologies and the search for involvement will concern most of the experiences that can be obtained during a holiday: the tourist offer should guarantee pleasant experiences through the use of special attractions such as virtual reality technology, pop-ups and entertainment for children. In light of the most recent changes, successful tour operators (also thanks to the support of new technologies) need no longer turn to individual consumers, but have to look at families as a whole, developing products offering increasingly personalised experiences.

Part 2
Technologies and the Sharing Economy in Tourism

3 NetGen and Tourism

Fabio Corbisiero

Introduction

The network society has brought about the birth of 'mass self-communication' (Castells, 2007) and is characterised by three elements: it is mass, as it is conveyed by the internet and peer-to-peer networks; it is multimodal, as relocation and content distribution; and, finally, it is autonomous in regard to the selection of the emission devices in many-to-many interactions. The communicative foundation of the network society, on the other hand, is built up by the global system of horizontal communication networks, which finds space above all in the second-generation internet, capable of connecting global and local. The network model, therefore, is flexible, adaptable and without borders. It is a centreless structure, based on interactivity, on the autonomy of each node and on the variability of scale. Each node has the same importance and is essential for the functioning of the network. The space–time dimensions are redefined: in fact, we speak of 'timeless time' and 'space of flows' (Bauman, 1998).

The so called 'Net society' (Castells, 1996; van Dijk, 1991) is a society in which a combination of social and media networks, information technologies and the internet is continuing to play a primary role. 'The diffusion of Internet, mobile communication, digital media, and a variety of tools of social software', Castells (2007: 246) writes, 'have prompted the development of horizontal networks of interactive communication that connect local and global in chosen time'.

The internet is one of the structural elements of post-modern time and contributes to creating new forms of sociality and community, based on multiplicity, plurality and the management of diversity, values and even the interests of people who use the technology. The experience of the Net has now become the terrain of a real information flow. When we talk about new information and communication technologies (ICTs), we generally no longer refer to traditional media, but to technologies such as computers, smartphones, tablets and many others, and their effectiveness and functionality through the internet. Unlike traditional media, ICTs allow interaction in a virtual space that becomes a place where everyone can not only dispose of information, but also produce and share it. From smartphones to social media, blogs to micro-blogs, Facebook to Face-Time, from TikTok to Instagram, the means through which we communicate, and the ways society thinks about communication are changing rapidly, perhaps more so than it did throughout the last century.

> The network society is also manifested in the transformation of sociability. Yet, what we observe is not the fading away of face to-face interaction or the increasing isolation of people in front of their computers. We know, from studies in different societies, that in most instances Internet users are more social, have more friends and contacts, and are more socially and politically active than non-users. Moreover, the more they use the Internet, the more they also engage in face-to-face interaction in all domains of their lives. Similarly, new forms of wireless communication, from mobile phone voice communication to SMSs, Wi-Fi and WiMax, substantially increase sociability, particularly for the younger groups of the population. The network society is a hyper-social society, not a society of isolation. People, by and large, do not fake their identity in the Internet, except for some teenagers experimenting with their lives. People fold the technology into their lives, link up virtual reality and real virtuality, they live in various technological forms of communication, articulating them as they need it. (Castells, 2005: 11)

With technology developing rapidly and exponentially, different technological applications are being used in tourism that are facilitating tourists' life. Specifically within the so-called 'robonomic society' (Ivanov, 2017), humanoid robots are playing an increasing role in tourism hospitality where artificial intelligence (AI) and automation technologies are used to improve the quality of leisure for tourists. Nowadays, you can text via the 'Four Seasons' mobile app for room service; 'Alexa for Hospitality' is being rolled out as an in-room concierge with some international hotels; and there is facial recognition on luxury cruise ships for quicker queues. The use of robots within the hospitality industry is becoming more commonplace, with uses ranging from artificially intelligent chatbots, designed to assist guests in any way possible, through to robot assistants, deployed to improve guests' experience in a hotel. Intelligent service robots are adopted mainly as robot concierges (e.g. Hilton's Connie) designed to deliver on-the-spot answers to questions, to suggest visit-worthy attractions and to self-learn for improved performance. Intelligent mobile robots, which self-navigate in indoor environments around people and objects, are used to transport items to hotel rooms, and pervasive devices (e.g. Wynn's Amazon Alexa), mostly operating on voice commands, assist hotel guests in controlling room ambience (temperature, lighting, sounding), making reservations, arranging laundry services and more. A further technological system relevant in tourism hospitality is the internet of things (IoT) (Atzori *et al.*, 2010), which is a (digital) network of physical objects, machine-to-machine communications or things that are wirelessly connected via smart sensors that allow tourists 'to communicate with several devices and services over the internet to achieve some useful objective' (Whitmore *et al.*, 2015: 267). Virtual reality (VR) enables individuals to have real experiences with the virtual world (Desai *et al.*, 2014), and augmented reality (AR) enables interaction between objects by combining the virtual and the real world. These applications of automation in tourism take place in tourists' daily lives and they are even useful to map the benefits and risks of a destination (Chung *et al.*, 2015: 589), although some applications are programmed to provide services and assistance before or after the travel journey. Intelligent automation, indeed, can be applied at the pre-trip stage to provide tourists with travel inspiration and assist them in the processes of information searching, booking and pre-arrival experiences. For service providers, deploying AI is critical for 'omni-channel' marketing automation to scale marketing content globally, to provide personalised offers and a more straightforward path to purchase for customers, and to generate new leads.

This change is now so fast that any empirical research performed at present is outdated long before it begins the process to publication. In this era of social media, ICTs have evolved from a broadcasting medium to a participatory platform that allows people to become the 'media' themselves for collaborating and sharing information (Li & Wang, 2011; Thevenot, 2007).

As already mentioned in Chapter 2, the rapid development of social media and the internet in particular has dramatically changed tourism processes. Being a mega trend that can significantly impact the tourism system (Corbisiero & Paura, 2020), ICTs and social media have been widely adopted by travellers to search, manage, share and write about their travel experiences through blogs (e.g. Blogger and Twitter), online communities (e.g. Facebook, RenRen and Tripadvisor), media sharing sites (e.g. Flickr and YouTube), social bookmarking sites (e.g. Delicious), social knowledge sharing sites (e.g. Wikitravel) and other tools in a collaborative way. New technological solutions are available for the real-time management of consumer relationships.

Social media and tourism are joining forces to make an impact on the way firms in the industry run their marketing campaigns. As consumer behaviour drastically changes, and visual and audio-visual content is reigning supreme on web platforms, hotel chains and other touristic players have exciting opportunities to target the people who could end up being their most satisfied guests. A few years ago, one-third of all leisure travellers in one of the most influential countries, the UK, chose their hotels on the basis of social media sites such as Tripadvisor and Facebook (Leung *et al.*, 2013). This is not surprising, considering that Web 2.0 is inherently a multimedia space, an aspect that according to Rokka (2010) transformed initially text-based research in tourism destinations into a visual approach based on the collection of pictures and visual as well as textual elements of Instagram posts. This exponential user growth in social media has changed the travel companies marketing strategies that use those platforms for product marketing through advertising, where they promote destination brands, discuss features and create social awareness. In addition, social networking sites are used to spread information faster than traditional news outlets or any other form of media. Everyday examples of social media sharing are seen through millions of photos that are uploaded to Instagram or other web places to share travel pictures.

The outbreak of Coronavirus disease (Covid-2019) has negatively affected our lives, increasing isolation and loneliness. Due to social distancing, travel and border restrictions, tourism is currently one of the most affected socioeconomic sectors (UNWTO, 2020b). However, the pandemic itself is transforming the proliferation of the internet and other technological innovations, affecting the way tourism destinations are perceived and consumed. The technological virtual world provides opportunities for destination marketing organisations to communicate with targeted markets by offering a rich environment for potential visitors to explore tourism destinations even during this dramatic era.

This chapter examines the relationship between generational change, technology and tourism, specifically focusing on the younger generations: millennials and Generation Z. We first highlight the use of digital technologies within the network society. We then discuss how digital technologies have influenced the travel behaviour of millennials and Generation Z to create their profiles, in order to generate social challenges and to help the tourism industry to create effective marketing strategies. Our main questions are: How is technology changing and influencing the link between new generations and tourism processes? What are young travellers up to? How do digital technologies impact young tourists' consumption experiences? The increasing use of technological devices by tourists has implications from a strategic standpoint; effective segmentation must be taken into account, as tourists' values may differ. Therefore, this chapter will specifically focus on these generations.

Technological Advances in Tourism

The nature of tourism is changing due to the growth of technologies. Nowadays, it is hard to imagine what travelling would be like without technology. Technological advances have changed the way people travel, and these new developments promise an even more interactive experience after the pandemic crisis. The contribution of new technological marketing to the promotion of tourist products, as well as the potential for tourism management and the process of decision-making are accelerating these processes. In recent years, some new 'mega trends' have emerged on the internet, underscoring changes that can significantly impact the tourism system. One of the biggest transformations that tourism has ever faced is the impact of ICTs, which have completely reshaped not only the way the whole industry works and destinations operate (Buhalis & Law, 2008), but also the tourist experience itself (Neuhofer et al., 2012; Tussyadiah & Fesenmaier, 2009). Particularly critical landmarks in this process have been the advent of the smartphone and its effects on tourists' experiences (Wang et al., 2014, 2016), online review sites and their influence on decision-making (Book et al., 2018) and search engines and the impact of social media on communication between tourists and on their behaviour (Bigné et al., 2016; Leung et al., 2013). Smartphones have become our tour guide, travel agency, best restaurant locator, map and more. According to Tripadvisor (2015), 45% of users use their smartphone for everything to do with their vacations and they have fundamentally reshaped the way tourism-related information is distributed and the way people plan for and consume travel (Buhalis & Law, 2008). Recent advancements such as cloud computing, sensors, GPS's widespread use, VR and AR, and the full adoption of social media and mobile technologies are continuously pushing forward the 'issue' of smartness in tourism (Xiang & Fesenmaier, 2017) to the point that the conceptualisation of destinations has evolved, and smart technology permits users to get all the pertinent information about their journey simply by using a phone, eliminating the need to download anything else (Ivars-Baidal et al., 2017). Importantly, these devices enable us to track not only those physical behaviours from the external information they provide but also we can guess quite accurately what travellers are thinking and how they are feeling at a specific moment (Swan, 2013).

Through the lens of the smart technology traveller, there are even more opportunities for tourism destinations to capture, understand and interpret contextual information generated by technologies connected to the internet. Social apps have expanded the aim of the tourism experience by enabling travellers to share their experiences with the rest of the world in different places whenever and wherever they want (Wang et al., 2014); besides the fact that wearable devices such as wristbands and smart iWatches and easily downloadable apps such as 'Master Tour' and 'Roadtrippers' have been widely adopted by consumers owing to their advantages of portability and potential usability for travel purposes.

Figure 3.1 illustrates how AI in tourism can be manifested in tourist-facing devices and interfaces (dark grey circle), powered by various AI capabilities (grey circle) to provide solutions to processes, functionalities, activities and experiences (light grey circle).

Those technologies seem to be moving travellers towards an increasingly data-driven sensor society (Choe & Fesenmaier, 2017) wherein an individual leaves a huge data footprint during the course of their everyday life, which creates opportunities for the tourist industry (e.g.

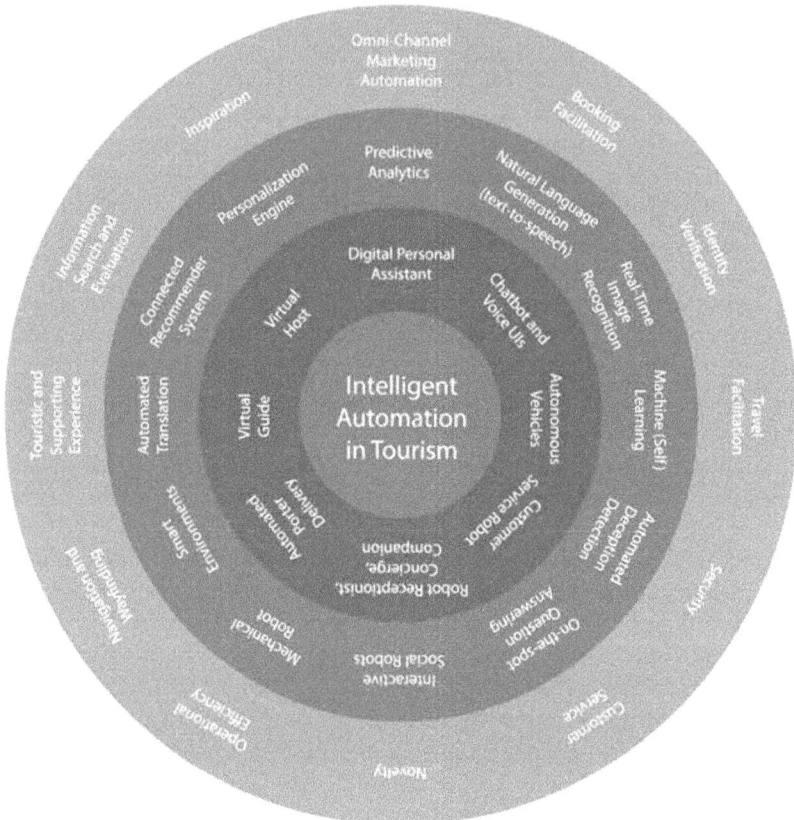

Figure 3.1 Intelligent automation in tourism (Source: Author's elaboration on Tussyadiah, 2020)

Andrejevic & Burdon, 2014; Swan, 2012). Information technologies have made enormous progress in terms of their ability to store and process large amounts of data, thanks to the evolution of cloud computing and data analysis tools. To all these ingredients we must add the young generations' propensity for digital data and their sharing through tools offered by the global world of social media networks. This new world of digital travelling is having a profound impact above all on young tourists' behaviour, business models and destinations at large (Femenia-Serra *et al.*, 2019). We are faced with 'smart tourism' which is shaped by technology and also by other factors such as new governance and management perspectives (Ivars-Baidal *et al.*, 2017) and a developing network of new forms of tourism (Gretzel *et al.*, 2015) or tourism on demand (Corbisiero & La Rocca, 2020). ICTs have generated a series of mutations in the dynamics and structure of tourism production and consumption both at a global level and as concerns tourism destinations, increasing 'competition' between generations and prompting tourism service providers to diversify their offer, on the one hand, and to understand and investigate the travellers' changing needs and requirements, on the other hand. As travelling becomes more digital and social, tourist providers are considering new business models to capture new generations of tourists, above all the younger generations. Multiple opportunities exist within this process, such as data-sharing arrangements to increase insight, exploiting competitor capability and having a full view of the tourist's journey. As such, many sources of big data that generate tourists' digital footprints become of great importance for scholars and practitioners to answer questions that are difficult to answer using conventional means such as tourist surveys and statistical data sources. For example, to identify leisure attractions for young people and to create tourist maps (Lin *et al.*, 2014), one can use geotagged photos from photo-sharing communities showing tourist attractions, which are rated as better than maps generated by similar methods and that are comparable to hand-designed tourist maps. Another way suggested by Chen *et al.* (2009) is to create tourist maps by clustering Flickr photos based on their locations and identifying the popular tags for those places.

The Hyper-Connected Generations

The homogenisation of technology and the resulting confluence of global consumers' tastes in travel have caused many to question whether age is genuinely a useful indicator of generational values and consumer preferences. It could be argued that the influence of global connectivity between different age groups, which has the effect of different generations gradually sharing similar habits and interests, is shaping social relationships even more strongly than social class differences. Today's generation of young people are more informed, more mobile and more adventurous than ever before. Youth tourism, defined as all tourism activities that are realised by young people aged between 15 and 29 (Horak & Weber, 2000), has become increasingly important since the UNWTO (2011a) estimated that around 20% of the 940 million international tourists travelling the world in 2010 were young people who generated $165 billion towards global tourism receipts and affirmed their financial value to the global tourism industry and local economies. Of course, that's not solely an economic dimension. International studies show that young people travel in order to experience a different culture, learn a language, volunteer, find a job or study, enhance relationships, escape and more (Khoshpakyants & Vidischcheva, 2012). Motivation has been suggested to be one of the most important dimensions for travel, especially for millennials who are keen to experience different lifestyles and meet new people during their trips (Obenour *et al.*, 2004). As the tourism industry itself goes through unprecedented changes, it is youth travel that has most contributed to such changes, through the innovation required by pioneering, heavy tech-using and socially and environmentally conscious tourists. While baby boomers grew up with radio, television and fax machines, while mobile phones and the internet were still developing, the Net Generation started to build their trips during the internet era (Skinner *et al.*, 2018; also Chapter 1). Raised alongside the first search engines, influenced by social media and constantly connected to their smart devices, these generations have grown up with a set of technicalities. Empowered by information and emboldened by their social impact, these social groups are poised to transform the way consumers throughout the world evaluate, purchase and consume goods and services. Technology plays a key role in the life of these groups of young tourists who tend to be tech-savvy, hyper-connected and internet addicts (CBI, 2019; Skinner *et al.*, 2018). Furthermore, these generations focus on experiences and the 'here and now' (Garikapati *et al.*, 2016) and choose to spend money on experiences such as travel rather than possessions (Cavagnaro *et al.*, 2018; see Chapter 4).

Sociological knowledge about the relationship between millennials' travelling behaviour and technology lags the attention of tourism marketers, which gives way to a number of concerns regarding permanent adaptation to consumer requirements. Notwithstanding an increasing interest in the millennial generation, existing research on youth tourism is relatively underdeveloped (Staffieri, 2016). For tourism marketers, identifying key travel benefits (and promoting those that are appropriate with respect to the destination) can be critical to the success of a travel destination (Migacz & Petrick, 2018). The appeal of millennials among tourism researchers and destination marketers is recognised as fundamental to analyse tourism in a generational way. Characterised as born 'wired' and defined as narcissistic, selfish and politically disengaged, millennials have more progressive attitudes and beliefs than do older generations on a wide range of issues, from the rights of sexual minorities to sustainable tourism itself, and many of them act on those beliefs engaging in green movements against touristification. Millennials disdain traditional marketing approaches and place a high level of trust in the suggestions of their peers and, more in general, of a wide spectrum of technological information sources (Belch & Belch, 2015). Furthermore, the values of this generation are embedded in 'micro-bubbles' – creative tourism, off-the-beaten-track tourism, alternative accommodation and fully digital tourism – that represent some of the key ways in which millennial travellers reshape supply and demand in the tourism industry.

These values affect destinations, attractions and other tourism businesses as they redefine what tourists want and how they want it. An IBM Institute for Business Value and the Economist Intelligence Unit research showed some aspects of this frame. The study surveyed 3017 travellers from 12 countries in 2014. To understand and predict the travel preferences and patterns of millennials, as well as to find out how they compare to previous generations, this survey reveals that for this generation technology is something intrinsic, which they apply to almost all aspects of their lives. Moreover, they significantly differ from previous generations in the way they value experiences, especially in their general disregard for wealth as a measure of success. They also exhibit a significant tendency to delay common milestones, such as marriage, home ownership and having children (see Chapters 4 and 5 for further details). Their aptitude for technology and new media means that millennials with these natural characteristics end up impacting business decision-making processes as well. Unlike baby boomers, millennials think that confrontation is essential to make decisions.

This scenario also inevitably involves the choices related to the purchasing process which for most millennials derive from teamwork, contrary to what baby boomers think, who in less than half of the cases consult their colleagues and are much less likely to share their travel experiences. The study also revealed the behavioural results of the generation of the most recurrent 'millennials' in the purchase phase of a technological device, which can be summarised here:

(1) when searching for products of interest, those belonging to the age group in question, which usually ranges from 18 to 35 years, tend to seek direct communication with the representatives of suppliers and colleagues, much more than Gen X or baby boomers, who are more likely to trust articles and blogs written by outside experts;
(2) after choosing the product, it is time for the second phase, the one that worries this generation the most, the convenience factor. Here in-person interactions become a burden rather than an advantage;
(3) In regard to the methods of communicating with suppliers, millennials mainly prefer email and telephone tools.

They are also prone to virtual meetings via video call and chat, and use WhatsApp and other instant messaging apps without any problems. But a strong factor is that communications have gradually moved more and more towards social media, considering that the number of young people in decision-making positions has grown: 81% of those in the 21–25 age group already used these tools to communicate, while it was only 18% in the previous generation. During the purchase phase, data analysis and personal contacts played a primary role – just as happened when making personal purchases. On the contrary, this research shows Gen Xs and baby boomers look to personal experience to make the right choice. This element can represent a real challenge for suppliers and manufacturers who need to rethink how their goods and services are perceived by potential buyers and also try to reach the personal networks of potential buyers. But the really important point of view is the social one. In this field, it is interesting to note that, according to the IBM study, it is very likely that millennials will increasingly tend to share their positive shopping experiences directly on suppliers' websites. They do not usually do the same, when, on the contrary, their expectations are disappointed: only less than 10% of millennials interviewed said they were willing to post a negative review on a company's website, on social profiles or on the website of third parties. Gen Xs and baby boomers are also very likely to share positive reviews across various channels, but have a greater propensity to share negative complaints.

However, even though millennials form a coherent group globally, there are also key variations across and within major world regions that we believe it is important to highlight. Indeed, we are talking about an age group that includes almost 1.8 billion people around the world (MSCI, 2020). Of the global 1.8 billion population, 1.1 billion or more than 60% reside in Asia, and 300 million or 16% reside in Africa. (Four of the five most populous countries in the world are in Asia – China, India, Indonesia and Pakistan – so this pattern should not surprise.) Of the remaining 400 million millennials, 150 million live in Europe and Latin America/Caribbean, with the rest in North America and Oceania (MSCI, 2020).

Confirming that millennials are practically always connected, this IBM report adds that 80% are very sensitive to online reviews, 57% upload photos several times a day, while 51% are glued to their smartphones to comment or post updates on their vacation. The choice of destination and bookings are often inspired by Facebook (54% of cases). In addition, 88% of users find the 'tips' from their contacts in terms of holidays useful. Not only that, but 83% clicked on the 'like' symbol at least once when they saw a brand linked to the world of travel on their wall.

The literature on the use of Facebook in tourism is very extensive, and, as detailed in Chapter 2, Instagram is also fundamental and has become a point of reference for those looking for inspiration for holidays. We are talking about a social network that records 150 million accesses a year aimed at collecting useful information on accommodation facilities and tourist locations. What about Twitter? It is also very strong: for 30% of travellers, hashtags – the compasses of the new millennium – become inseparable travel companions. Because they know that, between one tweet and another, there are also special offers and promotions from hotel chains or airlines. And still through hashtags, this generation looks for travel tips and advice on Twitter for tourism. If all this is not enough, the operators of the sector take into account another fact: not only do smartphones go on vacation with their owners (essential for accessing Facebook and various social networks), but 85% of them admit that they use it to plan their holidays, to book a hotel room (78%) or a plane flight (77%). It is no coincidence that 72% of those who use smartphones and tablets have downloaded at least one app dedicated to travel.

Also interesting is a further survey conducted by the luxury travel network Virtuoso (2015), which has revealed the secrets of luxury travel within the millennial generation. For this survey, even millennials, if carefully selected, can become excellent customers: travellers aged between 22 and 32 years tend, certainly, to use online agencies particularly frequently (87% of them do). And it is also true that their average daily travel expenditure is 62% lower than that of the 'mature' tourist segment. Nonetheless, if these potential customers are adequately engaged, it may be possible to significantly reduce this gap (down to 24%). These young globetrotters often prove to be more loyal customers than those belonging to other age groups. Virtuoso's analysis was carried out on the purchasing behaviour of tourists in the higher segment of the market based on their date of birth. The results obtained, while on the one hand undoubtedly confirm widespread opinions, on the other hand also investigate the reasons behind certain behaviours, at the same time denying some of the most consolidated clichés.

When analysing millennials, one has the feeling of being faced with a generation rather focused on itself, with little time available and stubbornly convinced that a smartphone can solve all of their problems. Tourism marketing must shape its proposals to this generation with ever-changing desires. For this reason, tourism service providers are warned: the priority for the future is also to model the sales channels based on the habits of millennials as consumers and to focus on simple and immediate communication, based on technology and social networks. If the tourism sector wants to prepare for the future by designing future-proof products and services, it has to take this generational change into account. Born into a digital age and with increasing international travel, this 'internet generation' (Özkan & Solmaz, 2015: 93) is likely to transform tourism and destinations just as much as the youngest one: Gen Z.

They are true digital natives, the first to grow up online, connected to vast amounts of timely, global information and instantly socially connected to the whole world. This generation grew up online and 'can get in contact with any person, in any location of the world, in seconds, and share information' (Berkup, 2014: 224). Using Tapscott's (2008) patterns to characterise this generation, we can argue that the criteria to describe their link with tourism are: customisation, freedom, scrutiny, speed, innovation and sustainability. Generation Z can also be described as individualist, consumerist, informed and digital (Tavares *et al.*, 2018) as shown by several studies.

A study across 45 countries has shown that around 60% of Gen Zers state that they feel more insecure without their mobile phone than without their wallet (Global Web Index, 2019: 5). This generation has grown up in an environment increasingly permeated by ICT although they are still developing their own values, habits, beliefs and financial independence (Carty, 2019). Gen Zers are multitaskers and expect to access and evaluate a broad range of information before purchasing (Wood, 2013). The combination of hyper-connection and critical global issues has made them concerned about humankind's impact on the planet and the future (Wunderman Thompson Commerce, 2019). If Generation Zers are frequent travellers they are at the same time aware and concerned about climate change and overtourism, and are willing to take a degree of responsibility in mitigating these phenomena by making better booking and travel decisions. Gen Zers have, indeed, been portrayed as socially and environmentally conscious, wanting a mobile-first approach and desiring authentic local experiences (Bec *et al.*, 2019). A Generation Z tourism research experience in New Zealand shows, once again, that Generation Z is characterised as 'being digitally adept, social and mobile, ICT is a critical component and a linkage to services or to the outer world' (Robinson & Schänzel, 2019: 94). The widespread use of mobile apps, such as 'Campmate', and the reliance on social media are a common feature among the participants in this research. Technological advancements facilitate ease of access to information, facilities and places. Therefore, Generation Z's behavioural patterns in a destination and their experiences may be impacted or influenced by ICT advancements. The European Travel Commission (ETC, 2020) collected responses from respondents of this generation who reside in China, Germany, the UK and the US. Delving into the relationship between Generation Z, tourism and technology, this study focuses on some interesting aspects of Generation Z. First of all, staying in contact online and sharing experiences in real time play a substantial part in Gen Zers' travel habits. More than 40% of the respondents in each country stated that they regularly engage in instant messaging or posting live photos and videos. Interestingly, while Chinese Gen Zers prefer to share photos or videos in real time, their counterparts from the US and the UK are more inclined to share their experiences after the trip; this is possibly because networks, such as Instagram (highly popular in the UK and US markets) require slightly more time for photo curation and editing. While some travellers have reacted to accelerating digitalisation by seeking occasional disconnection while travelling, this trend does not appear to apply to Gen Zers from China, the UK and the US. German Gen Zers showed slightly more willingness to disconnect while travelling (31%), while Chinese Gen Zers were the least likely to disconnect (17%). It was noted in this research that Generation Z is a generation for which the mobile phone is the most important device, and this can be seen in their travel habits. Related to the previous recommendation that national tourism organisations (NTOs) and private business should foster the creation of mobile-friendly platforms, these findings suggest that it is equally important that NTOs curate experiences for Gen Zers that are easily sharable and 'Instagrammable'. Additionally, just over 40% of Gen Zers in each country agree that flights should be taxed if not endowed with green technologies or that the number of flights should be reduced. Despite travelling being an important aspect of the life of Gen Zers, introducing additional taxes which will eventually make prices higher would be seen by them as a reasonable decision to be taken in the context of measures to combat climate change and reduce the ecological footprint. Youth and youthful travellers, demographic change and state-of-the-art technological devices represent an important phenomenon which may pose both opportunities and challenges for the development of tourism and destinations. Some scholars (e.g. Gardiner *et al.*, 2014) indicate that future travel behaviour will differ between the generations. Therefore, there is an implied necessity for continuous studies and research on each generation in order to effectively respond to the needs and demands of each of them. Generation-based research that identifies different groups of consumers and their specific needs and desires is therefore important (Chhetri *et al.*, 2014). Recent findings, for instance, indicate that less technologically enabled tourism destinations can benefit by employing contemporary principles and practices to meet the needs of the new generation of tourists who seek rich digital and gamified tourism experiences (Skinner *et al.*, 2018). Both millennials and Gen Zers consider contemporary tourism as a social and cultural experience that encourages socialisation and identity construction within a growing interest in new tourism practices and niche proposals, giving new meaning to their choices as tourists (Monaco, 2018). Being two generations that grew up in times of unprecedented connectivity and ease of travel, millennials and Gen Zers have come to believe that travelling is a necessity for human beings and that people have a 'right of mobility' (Monaco, 2019b). They also realise that tourism can have damaging effects, and this can be seen in the universal conviction that too many tourists can be detrimental to locals. In this perspective, one can expect to see more destinations and attractions developing unique products to cater for generational tourism; new destinations emerging on the tourism map with the help of the creative trends of younger travellers with the never-ending growth of digital tourism platforms and apps. However, the smartness of destinations will not be conditioned exclusively by technological tools and new generations. It is clear that a strategy that is shared by all actors in a destination will be essential in order to take full advantage of the opportunities generated by the technological evolution, with solutions adapted to each territorial and tourism context. There is much scope for further social research on these important travel generations. Specifically, there is a need to figure out how travel motivations and behaviours change over the course of the travel paths. The social researcher should indicate that different cohorts of young people have behaved differently over time, but more longitudinal research is needed to confirm this. One possibility is to develop travel biographies for different generations of travellers, and to use qualitative and quantitative data to reconstruct a world history of youth travel over recent decades. This might also help us understand the links between the wide social, economic and geographic contexts of travel and the behaviour of young globetrotters.

Conclusion

In this chapter, we examined how significant the interconnections between tourism and Net Generations are. In particular, we highlighted how and in what regard millennials and Gen Zers are using new technologies to disengage forms of mass tourism and engage their interests in more sustainable forms of tourism.

Digital technologies have reduced the lines between the real and the virtual world, offering the potential to increase the level of more sustainable involvement within tourism experiences. Specifically, artificial intelligence, VR, AR and social media have the capacity to enhance visitor experiences both at tourism destinations and prior to visiting tourism destinations, and importantly for recollecting tourism destinations (Little *et al.*, 2018). The habits and needs of the contemporary traveller are much different from those of the past. Millennials and Gen Zers in particular are constantly looking for 'conscious tourism' and experiences that are able to combine innovation and authenticity in the best possible way. This more conscious involvement within tourism experiences plays a more critical role in helping tourist destinations achieve more sustainable hospitality. These young tourists act precisely with a view to authenticity that, for example, transformative travel experiences have multiplied, or trips dedicated to personal growth where it is possible to develop specific skills or do cultural exchanges or volunteering. In fact, on the one hand, a large slice of travellers is interested in experiencing increasingly high-performance travel experiences in the name of hi-tech. On the other hand, however, there are also those who take advantage of the journey precisely to switch off and disconnect from the rest of the world, at least for the duration of a vacation.

Until the advent of ICT and social media the characteristics of travel were limited by physical spatial constraints and the related time. The characteristics of tourism depended above all on the traveller rooted in a limited area. The development of the internet and the birth of social media have made it possible to share the journey, its experiences and its feelings with all the visitors of the web. In particular, by allowing a permanent connection with peer social networks, digital media are becoming a key resource for the construction of a tourist identity that is largely influenced no longer by the real 'tourist gaze' (Urry & Larsen, 2011) but by that mediated by the distance of social networks. The 'kodakisation' of the journey has created an 'always on' tourist, who uses their devices to share tourism experiences on the spot and in real time.

Displaying, identifying, searching and sharing tourism experiences and information are identified as the top two ways in which new generations and social media have transformed tourism (Corbisiero, 2020). Overall, social media have significantly changed the way young people not only consume but also produce and communicate/share travel information, which in turn influences the way young tourists both select and experience destinations. The syndrome that 'everyone is watching me' has also changed the way young tourists act in tourist places in terms of what they see and perform at a destination, how they interact with landmarks, what they are 'forced' to externalise and communicate as 'socially desired' and self-enhancing emotions and touristic behaviours. Unlike the traditional photo album, exhibitions of pictures on Instagram and Facebook or video on TikTok are tied into the flow of the everyday and tend to highlight a culture of suddenness where people expect instant delivery, ubiquitous availability and gratification. In the quest for self-promotion and the search of an idealised tourism experience, young tourists create a collective form of traveller with the direct aim of widening their popularity and the latent one of widening the popularity of the tourist spaces and places from which their 'digital action' starts.

Due to Covid-19, the number of Europeans travelling outside of Europe decreased by 98% in June 2020 compared to June 2019 (Eurostat, 2020). Net generation travel outside of Europe has almost completely vanished. However, this target group promises opportunities for the future. According to the GlobalData (2020) survey, young travellers will be vital to the recovery of the travel and tourism sector. These young travellers are likely the first ones to take international trips once the restrictions have been lifted. While the future at large is still uncertain, what is certain is that the globalising impact of the pandemic has fuelled a disproportionate use of new technologies that, in turn, has accelerated the digitisation process of old and new generations. We are currently facing a 'tourism of the minds' which has temporarily replaced the 'tourism of the bodies'. Some mega trend papers published during the pandemic (Corbisiero, 2020; Corbisiero & Paura, 2020) describe youth tourism as the major source to discover how tourism will be after the pandemic. As trendsetters, youth travellers make previously undiscovered destinations accessible and sustainable. For many young travellers, the pent-up demand to travel caused by a lack of it due to lockdown will have encouraged a greater need to experience a new feeling with a tourism experience. Recent research by Topdeck Travel (2020) revealed that 28% of Gen Z surveyed admit that they will seek meaningful and authentic travel experiences in the very near future, and 23% are hoping to experience a different culture to what they have been used to before the spread of Covid-19. Hoping for a stronger tourism growth rebound with a recovery to 2019 levels by 2023, the tourism gaze will return to watch on the world. The future of the travel experience will be a seamless blend of young travellers and new technology where machines are tasked to do more of the 'machine' work, freeing and empowering destinations to provide better service experiences and where youth generations are tasked to create more meaningful social connections.

After several lockdowns in all the countries of the world, young people will be desiring to escape from their everyday lives at home. It is likely that young travellers will want to meet new people, gain new perspectives and create meaningful memories through social media and social networks. These developments are changing rapidly, and they need to be closely monitored.

4 Towards a Sharing Generation

Salvatore Monaco

Introduction

From a sociological point of view, it is interesting to attest that at a global level young people have gradually become protagonists of a way of consuming and experiencing that seems to partially depart from the most traditional market rules. In other words, in the post-modern society they are creating unprecedented forms of value. In particular, several studies have highlighted a strong predisposition demonstrated by members of the younger generations to question the typical relationships between the parts that characterise the market. Thus, they are implementing new forms of consumption, defined in the literature as 'collaborative'. The term was first used in US scientific production in the late 1970s (Felson & Spaeth, 1978), and research on the topic has intensified and strengthened since 2010 (Algar, 2007; Hamari *et al*., 2015; Rogers & Botsman, 2010; Tuttle, 2014).

According to the anthropological and sociological literature (e.g. Gouldner, 1960; Simmel, 1950; Westermarck, 1908), the exchange not only of material but also symbolic resources has allowed the evolution and the survival of primitive man and it has continued to the present day as a principle of social cohesion. Reciprocity is a subject on which the social sciences have much argued, so much so that some scholars have even spoken of 'homo reciprocus' (Becker, 1956).

The sharing economy emerged in response to the economic crisis of 2008–2011. The diffusion of technologies and the use of collaborative consumption platforms have accentuated, consolidated and made much more visible a post-modern version of reciprocity, that is, an economic model focusing not on the possession of things, but on sharing them. In other words, this system is based less on the purchasing of goods and services and more on their use (Smith, 2016). In this scenario, the various online platforms, acting as digital intermediaries, over the course of a decade, have contributed to the spread of a very heterogeneous offer of services which have enabled, also at the tourism level, peer exchange practices (Schor & Fitzmaurice, 2015), extracting value from the resources made available by users. Botsman (2017) argued that, through their behaviours, the new generations have implicitly introduced at the social level a 'culture of the us', which has supported (and has partially overcome) a more individualistic way of thinking and acting, characterising the behaviour of previous generations. The author believes that the diffusion of sharing has been fast thanks to the presence of technology, which has made collaboration possible through a permanent network connection. The main difference between post-modern reciprocity and that of the past is represented by the network within which gifts, exchanges and loans are no longer restricted to friends, relatives, neighbours and acquaintances, but is on a global scale. In fact, 'the increasing connectivity, propagated by online social network platforms, allows people to share access to products and services among each other' (Tussyadiah & Inversini, 2015: 817).

Online platforms allow users to settle on a series of agreements, which in most cases people do not know each other personally, but have similar goals in common. This phenomenon, however, was defined recently in an article published in *Forbes* as 'NOwnership' (Morgan, 2019), underlining the less strong interest shown by members of the younger generations in the ownership of various consumer goods. This trend prompted jurists Aaron Perzanowski and Jason Schultz (2016) to speak of 'end of ownership'. In fact, nowadays there is a subscription, which can be shared with other people (who know each other or not), for everything: from sport to cinema on demand. Tien Tzuo, chief executive officer (CEO) of Zuora, defined it as the 'subscription economy' (Tzuo & Weisert, 2018); it no longer matters what a person owns, but what a person does, through a series of experiences. In other words, people increasingly prefer to subscribe or rent than purchase.

In the tourism sector, the most common sharing practices include the sharing of means of transport, the temporary exchange of homes, the free stay at the home of someone who makes available a space in their home and the loan of material goods. Other interesting sharing practices are team purchases where individual consumers are able to enjoy the best discounts or special offers originally reserved for 'teams' made up of at least two people by joining the existing 'team' or calling on their social network to form one.

Sharing Car and Home

The transition from the logic of ownership to that of access shows the different value that today's young people attribute to some material goods, which previous generations looked upon as real status symbols, but in contemporary society do not fall into the category of priority purchases. Typical examples can be found when it comes to a car and a home. For a long time, people thought that these goods had to be purchased, and also to have concrete things that testified their realisation in life and their social position (e.g. Belsky *et al.*, 2014; Zinola, 2018). Until a few years ago, the purchase of a home was seen as an investment, an object of value to be inherited by children and grandchildren in the years that followed. From millennials onwards this type of approach appears almost obsolete: buying items has taken on a very different value from what it had for their parents and grandparents. The Nielsen Global Generational Lifestyles Survey (Nielsen, 2019) revealed that only 22% of millennials consider homeownership a top priority. Data were collected on 30,000 online respondents in 60 countries throughout Asia-Pacific, Europe, Latin America, the Middle East/Africa and North America. Similarly, research in America has shown that renting a house, a long considered temporary solution, is silently emerging as a new lifestyle choice (e.g. Derber, 2015; Dunn, 2019; Sengupta, 2017). But renting is not the only option. The Danish company 'Space 10', an 'Ikea spin-off', and the American company 'Anton and Irene' have launched the 'One Shared House 2030' project, proposing co-living solutions that to date have already involved more than 80,000 people all over the world.

In Eastern culture, however, the scenario of apartment sharing differs from that of the Western world. For example, influenced by traditional Confucian philosophy, Chinese people still hold traditional Chinese concepts such as 'house precedes family', 'house is the necessary condition of a marriage' and 'a self-owned house would bring happiness and a sense of belonging'. Ioannides and Rosenthal (1995) contend that consumers who think that a house brings them strong satisfaction and a sense of belonging are more likely to buy a house. Wang (2016) has conducted a questionnaire survey with a sample of 87 respondents of which 97.7% are millennials. Forty respondents choose to rent a house as a temporary strategy with the intention to buy one in the future. The survey also shows that, in line with the housing gradient consumption theory, urban residents make a rational choice when they buy or rent a house on the basis of their objective conditions such as financial income. Economic difference gives rise to a common makeshift strategy of renting and buying a house in China: new graduates or white-collar workers with a middle income, more often than not, would rent an apartment for a period of time before buying a small apartment with some financial reserve; then, right after their capital is accumulated, combined with money from selling or renting out the small apartment, they would ultimately buy and settle down in a bigger apartment. Also, Wang calls for the government to formulate laws and regulations regarding the house rental market, encouraging real estate developers and investors to afford a stable supply of houses for rent, thereby favouring its benign development. Recent years have witnessed a dramatic change not only in the mindset of young generations of Chinese people but also, with the assistance of technology, in the niche of apartment rental. According to the recent survey published in the 2019 White Paper of China's

House-Renting Industry Security, a total of 160,000,000 Chinese choose to rent an apartment in Chinese cities and towns, accounting for 21% of the total permanent resident population. Newly employed university students are the main group of residents. An analysis of the objective demand and supply relationship in this sector demonstrates the following facts: (1) the Chinese government has successfully taken action to control the population in major cities and stabilise the property market. The policy of restricting house buying, especially in metropolitan cities such as Beijing and Shanghai, will continue to promote the development of the house-renting market. (2) Against the backdrop of consumption upgrading, an increasing number of young people are pursuing a quality of life; rather than buying a house they plan to rent in the long term. (3) By the end of 2017, China had a migrant population of 244,000,000 and 240 million single people, which led to higher demand in the rental market. (4) By 2017, the apartment vacancy rate in China had reached 21.4%, and most proprietors were willing to rent their property as an investment. The skyrocketing house prices in China are a principal factor in determining people's choice over buying or renting: the price-to-rent ratio in China's first-tier cities (such as Beijing and Shanghai) has reached 500:1, that is, a total of 500 months of the average salary of a Chinese person can cover the investment in his or her own apartment. On the international scale, however, this ratio is 250:1, one half that of China. The ratio shows the house-buying craze of Chinese people on the one hand, and a high price/performance on the other. As a result, middle-aged Chinese immerse themselves in real estate speculation, while the Generation Y are awakening, as they are increasingly influenced by concepts introduced from the Western world, realising that buying an apartment is not an end in itself. They are starting to refuse to be a 'slave' to a mortgage, a popular term in China to describe those who spend almost all of their monthly salary to pay a mortgage but have a lower standard of life. In response to this change in the house rental market in China, a large number of companies or intermediaries offering short-term and long-term leases for white collar workers and new graduates have sprung up in recent years. ZiRoom is a giant in this sector: it has adopted an O2O model in the long-term apartment rental market, that is, a new business model in which retailers use both online and offline channels as an intensive strategy. The company provides nearly 500,000 apartments for rent and serves more than 3,000,000 tenants in nine major cities in China. It has launched several services to help tenants feel at ease (just as its company name suggests in Chinese): completely furnished, fashionably decorated, online reservation for furniture maintenance on an app, regular house cleaning (also with the support of a personal housekeeper who provides cleaning and other household services) and so on. In cooperation with JD Finance, the financial subsidiary of China's second biggest e-commerce giant JD.com Inc, ZiRoom has launched an internet credit payment service

'Baitiao' in the model of 'buy now, pay later', that is, in instalments. This service greatly helps, in particular, the ever-increasing number of new graduates (the figure reached an all-time high in 2020: 8,740,000) and the white-collar millennials who have too low an income to afford to rent a comfortable place. Another quintessential example is Danke Apartment. Phoenix Tree Holdings, or Danke Apartment, has established its stable house resources in 13 major cities across China with a total of 406,746 apartments as of 30 September 2019. Unlike traditional C2C rental intermediary, Denke adopts a C2B2C business model in staging its internet-based business, centralising the apartments and standardising its operation, thus becoming one of the biggest and fastest-growing co-living service providers in the country. On 17 January 2020, the company was listed on the US initial public offerings (IPO) market under the ticker symbol of 'DNK' with 9.6 million American depositary shares (ADS). In the first half of 2020, Danke cooperated with 12 different companies that offer life services covering cosmetics, food and beverage, healthcare, online shopping, entertainment and freight transport with the aim of ensuring an all-round service for its tenants. For example, it has launched a joint project with Shunfeng, one of the famous courier service companies in China, to help new graduates with their luggage. With only a click of a mouse on a PC or a few taps on a smartphone, they will be granted postage coupons and can reserve a pick-up within one to two hours. In regard to leisure activities, for the office worker who has a preferences for hiking in the suburbs on weekends, Danke also has a car rental service on its app in cooperation with a car rental company. 'Get coupon first and then make a reservation' makes it possible for people to travel at a lower cost. Any 'credit' on the app helps tenants to rent a better model or enjoy a better discount. The appearance of various apartment rental companies such as ZiRoom and Denke has not only helped ease, to varying degrees, the economic burden and difficulties of finding a suitable apartment, but it has also addressed millennials' need for socialisation. In an attempt to satisfy the emotional needs of youngsters, ZiRoom has also launched a service product called ZiRoom Quality Community in Beijing, Shenzhen and Shanghai, offering tenants different social activities including café parties, football matches, parkour battles, cuisine meetings and cosmetic salons. All these activities cater to the different needs of tenants. In the tourism sector, however, ZiRoom has also launched its ZiRoom Minshuku (homestay product). Tourists are able to contact the landlord and confirm the location on its user-friendly app. With its determination to offer the most beautiful homestay houses in China, ZiRoom has recruited professional interior designers and artists. In 2016, ZiRoom carried out a crossover cooperation with the designer Shen Hong in a project to redesign and renovate an antic villa based in Shudefang, Shanghai, with a history of more than 80 years. The building is not merely an ordinary garden villa in Shanghai, but it represents the true life of Shanghai people and their underlying culture and customs. Such a transformation marks a step forward to modern times in both construction and lifestyle, giving the tourists who stay in the villa an illusion of reading the past by touching the old bricks while also enjoying the facility of modern technologies.

Similarly, nowadays the 'car obsession' (Cross, 2018) is disappearing. Recent studies (Sivak & Schoettle, 2016) claim that in many countries the number of new licenses has fallen by 20–30% compared to that of 2001 and the proportion of car purchases by people under the age of 30 has decreased from 15% in 2008 to 8% in 2017. Globally, the empathic relationship that an 18 year old had with the car until a few years ago has practically crumbled (Bain and Company, 2019), because of the costs incurred, not only for the purchase of the vehicle, but also potentially hidden costs such as parking, insurance and maintenance. Instead of buying a car, young people are more inclined towards car sharing, ride sharing and carpooling. In fact, for most under 35s it is not important to own a car, but to have one available when needed. The car is no longer a priority: young people think of a car as something useful, but not essential, just a simple means of transport, little more than a commodity (Pilia, 2019). The preferences and behaviours of young consumers have already triggered a profound change in the automotive industry, so much so that economist Jeremy Rifkin in 2011 argued that within 25 years the norm would become car sharing and the anomaly would, in reverse, become car ownership. The most forward-looking companies are already anticipating these future market trends. In particular, some of the largest car manufacturers, from Mercedes to General Motors, from FCA to the Chinese Lynk and Co., are promoting car sharing initiatives and are producing cars ready prepared as standard for sharing, designed primarily for younger people.

Shared mobility services are various: scooter rental; bike sharing; station-based car sharing in which a car is picked up and left in special spaces; and free floating car sharing, a formula that allows people to rent cars and park them anywhere within the perimeter allowed by the operator. Finally, there is the sharing of personal means of transport. Peer-to-peer sharing takes place when a private user makes his or her car available to other people (such as friends, colleagues or members of a dedicated social network): in theory, they all make a profit and free up space in the cities.

In North America, car sharing services were introduced in Quebec City in 1994 by Benoît Robert and a company called Communauto which is still a leader in car rental globally. Car sharing practices started to attract the attention of American people in the 2000s, thanks to the presence on the market of various players who launched more and more personalised products and services, in line with user expectations. The success of Zipcar, Flexcar (purchased by Zipcar in 2007) and City Car

Club stimulated several car rental companies to launch car sharing services in the US and in many European countries. Among the various services are Avis on Location by Avis, Hertz on Demand (previously known as Connect by Hertz), Uhaul Car Share owned by U-Haul and WeCar by Enterprise Rent-A-Car. During the first decade of the 2000s there was a total of 730,000 members globally sharing a total of 11,000 vehicles (Chan & Shaheen, 2012). Car sharing has gradually spread to other global markets, establishing itself primarily in urban areas characterised by dense populations, overall in Argentina, Brazil, China, India, Mexico, Russia and Turkey. In 2018, the main cities in the world for car sharing were Tokyo, Moscow, Beijing and Shanghai (Shaheen *et al.*, 2015).

Many countries sponsor car sharing practices with the aim of promoting a new social and economic model to respond intelligently to contemporary ecological challenges. In this sense, the UK is among the most virtuous countries in Europe for the efforts put in place to reduce CO_2 emissions taking advantage of sharing services.

There, car sharing is promoted by Carplus, which is a national charity whose mission is to encourage the responsible use of cars in order to alleviate the financial, environmental and social costs of motoring. Carplus is supported by Transport for London with the British government's initiative to reduce congestion and parking pressure and help ease the load burden on the environment and reduce traffic-related air pollution in London.

Looking at the Asian world, also in China there is a downward trend in young people's desire to buy a car. For the first time in history, negative growth of China's total volume of auto retail sales was seen in 2018. The following year also marked a decrease of 3.4% year on year. In view of this crisis, several studies claim that the automobile market has been experiencing a 'severely cold winter'. The reasons behind this phenomenon are varied: in the first place, most Chinese families are feeling increasingly stressed with the normal spending on house and children. Their economic difficulties may be compounded by the two-child policy (a policy that allows every Chinese family to have two children with the aim of guarding against the increasingly aging Chinese population), let alone the cost of buying a car and other related costs of maintenance and insurance. Secondly, metropolitan cities such as Beijing ban automobiles with even- and odd-numbered license plates on certain days. Some of China's municipalities such as Tianjin and Beijing restrict traffic flow and issue car license plates via bidding or a lottery in an attempt to fight against traffic congestion and reduce air pollution. As a result, young people find it difficult to get a license plate, and even if they were determined to get one, they would probably wait for months or years, which greatly discourages them from buying or owning a car. Last but not least, with the introduction of the shared economy and the launch of various online car rental apps in China, young people are more willing to 'rent' a car than own one, because owning a car does not guarantee the satisfaction or happiness they expected, instead, probably adding more economic burden or the anxiety of being unable to use it on days of license plate number prohibition. The starting year of China's shared automobile was 2015. From this year onward, the regional pilot scheme of shared automobiles extended from several small companies in North China to more than 200 enterprises across China. With the introduction of the new concepts 'new-energy automobile' and 'ride sharing', many talents and investors have been attracted by this new industry. In 2017, Didi (a Chinese platform that offers a one-stop mobility service) announced its cooperation with the Global Energy Interconnection Development and Cooperation Organisation (GEIDCO) in founding a new global company of new energy automobiles and it also plans to launch over 1 million new-energy cars for rent on its platform. At the end of 2017 and the beginning of 2018, both Mobai and Xiechen, another two big companies that serve mobility in China officially launched the rental service of shared automobiles, not only in metropolitan cities but also in neighbouring cities where residents have travel demands. According to iResearch (2020), as of February 2019, more than 1,600 companies involving shared automobile have been registered in the Chinese market; 110,000–130,000 automobiles for rent have been put into the market, generating a giant market size worth 2,850 million RMB. Long gone are the days in the 1980s when owning a bike with the label of Yongjiu (meaning 'forever') was considered as having a luxury good. With the implementation of the policy of reform and opening up, buying a bike is now common and affordable to most Chinese in the contemporary era. As early as 2007, the model of bike sharing was introduced in China and was administered by the urban government. Since 2014, however, as the mobile internet has rapidly grown, internet-based bike sharing has become popular. Users are able to look for the nearest available bike and unlock it on an app and are charged only 1 Yuan RMB per hour. A survey shows that bike sharing attracts users in the 25–35 demographic who ride a bike three or four times a week. In 2016, at the peak of development, at last 25 new brands joined the market with Mobike and Ofo the biggest leaders. Riding a shared bike is seen not only as eco-friendly transportation but also a fashionable behaviour. Chinese pop stars Luhan and Liu Haoran have both endorsed the shared bike products of Ofo and Didi, which attracts millions of their fans and youngsters to follow suit and helps, at the same time, to promote the idea of eco-friendly commuting.

However, worldwide not all countries are aiding carpooling. For example, in Hungary, a car offered to someone in exchange for compensation is seen as a tax offence unless the driver has a taxi driver's license, an invoice is issued and the taxes are paid at the end of the service. As for Italy, in 2018 more than 5 million Italians used forms of shared mobility, amounting to year-on-year growth of 25%. In Italy, the various car sharing, scooter sharing, carpooling and bike sharing services have arrived, with 363 present in

271 municipalities, 57% of which are located in the northern part of the country, with Milan confirmed as the best served city (ONSM, 2019). In more recent times, FCA Bank and its subsidiary Leasys have launched a new car subscription service in Italy called CarCloud. CarCloud allows people to rent a car in the main Italian cities, seven models (from 500 to Jeep Compass) from which customers can choose, paying a subscription fee with a monthly range of 1,500 km. All operations, from registration to payment, up to the choice of car are made available online: the activation of the service would firstly need the registration of an Amazon account that enables the user to choose between the two available packages (from 249 euros per month for a Fiat 500 to 349 euros for a jeep, plus registration fee). Stamp tax, insurance, maintenance and tyre change are included in the package. Customers can collect the car in 1 of 150 affiliated stores in Italy.

Sharing to Save

One of the most obvious advantages of collaborative consumption is that this kind of activity would guarantee people the chance of touring even under the trying circumstances of a credit crisis. The main feature of sharing activities is that they allow people to generate experiences at low costs as a result of disintermediation. Specific collaboration platforms on which people share experiences provide users with the opportunity to get in touch with each other, bypassing some intermediary players in the tourism chain.

This sociological interpretation of the phenomenon allows to maintain that collaborative consumption responds to the social need to reduce differences in access to tourism, allowing the implementation of an alternative redistributive model: the idea of sharing gives rise to a possibility that can deal with inequality – at least in part and in specific areas.

Recent studies (e.g. Kurz *et al.*, 2018) pointed out that American millennials have low income and wealth compared to members of other generations (baby boomers and Generation X) when they were at their age. In fact, while previous generations lived their 20 years in a period of growing per capita gross domestic product (GDP), millennials approached adult life with a reduced per capita GDP, with the consequence that contemporary young people have less disposable income compared to Generation X. The recession has left lasting effects, directing younger generations towards the same consumer goods as previous generations, but with an extra focus on the economic aspects. In a recent study, 94% of US millennials (a sample of 500 individuals) said that they had made use of the sharing economy at least once, from Airbnb to Uber (Lab42, 2019). The research confirms the aptitude of new generations to seek immediate, efficient and varied consumption. For this reason, a sub-segment of the sharing economy, the 'rental economy', is growing increasingly strong: the object of desire, be it a television series, a work of art, a luxury handbag or the latest album of a rock band, can be rented. In doing so, young people get what they want, spend less and abandon the object of desire when it is no longer of interest to them. According to the study, the consumer goods most hired by American millennials are furniture. Ikea has already considered experimenting with some subscription leasing programmes in 30 countries. The moves of other players in the furniture market will depend on the success of this experiment. Videogames, professional work tools and clothes follow. More specifically, clothes are hired mainly for special occasions, both formal and ceremonial (77%) and professional (54%). The study also highlights some gender-related differences. Women dominate in the rental of fashion and furnishings, while men in the rental of video games and DIY tools. As for clothes and accessories, the service most used by young Americans is 'Rent the Runway', which shares the market with many other companies, mainly start-ups, which are popping up like mushrooms, such as 'Nuuly'. This company offers the opportunity to rent clothes with over 100 vintage brands to choose from, charging $88 per month. Other similar services, at more competitive prices, are capturing the attention especially of younger people and students, who sometimes have tight budgets available for shopping. This is the case, for example, of 'American Eagle's Style Drop', a company that offers its members the opportunity to rent clothes, charging a monthly payment of approximately $50. Clearly, the behaviour of young people in the US is conditioned by the American social, economic and political context. However, from a forecasting point of view, it is possible to imagine that the attitudes of young Americans will be replicated in the medium to long term also in Europe.

The rental economy, although at a preliminary stage, has good potential for growth, so the global retail giants that want to attract the attention of future generations should perhaps consider changing their business models now.

As for the Eastern world, the concept of wardrobe sharing has become a favoured choice of many young Chinese women. College students and white collar workers are the major customers. Online platforms of shared clothing such as Ycloset, E-Cool and Ms Paris enable their users to rent different clothes for an unlimited time. Ycloset stands out most with several rounds of investment and a strategic relationship with LVMH. According to the advertising of Ycloset, they boast more than 1,000,000 pieces of fashion clothes covering international designer brands such as Kenzo, Pinko, Self-portrait and other hot brands. Users pay only 499 Yuan RMB a month for different rentals over an unlimited time period; three pieces at a time. It is mostly gown styles worn on limited special occasions that are rented by users. It has launched the idea of a 'virtual wardrobe' where users gain access to all pieces by paying the membership fee; or if they wish to own an item after having rented it, they are able to buy it, which is manifested in the idea of 'experience consumption'. As shown by the report of Ycloset, their clients are mostly women aged around 28 years old, that is, millennial women. It is the age between that of a new graduate who has little knowledge of upgrading their wardrobe and the age of a mature white collar worker who has better consuming power. In an attempt to help women at this age with their wish for better clothes and less consumption, Ycloset has upgraded its system of 'stylist plus algorithm' which automatically offers according to the user behaviour of every client and it has also established multiple warehouses equipped with the function of clothes washing, sanitising and packing across China. In the first half of 2020, Xianyu, a second-hand marketplace affiliated to China's e-commerce giant Alibaba, also started its rental of various categories; people can trade and rent both used and unused items on this marketplace; with just a few clicks or taps, the original item purchased on Taobao (one of the largest e-commerce platform in China) can be sold or rented within a few hours.

In the tourism sector, by investigating in detail the drivers that guide collaborative consumption, it is possible to argue that the economic dimension, which probably represents the most expected and foreseeable one, is accompanied by two other vectors, which are sustainability and the relational dimension. As we will see in the following pages, these two drivers play a central role in the same way as the economic dimension.

The Sustainable Approach to Consumption

As anticipated, in these unprecedented times of consumption, the economic aspect appears complementary to other drivers. A second important aspect that feeds collaborative consumption is sustainability: as we have already seen, younger generations are much more environmentally conscious than previous generations when it comes to issues related to the protection of the eco-system. A few years ago, *The Economist* (2017) published an online article explaining that millennials are the generation most involved in so-called socially responsible investments (SRI). In fact, millennials think that economic and financial choices cannot be separated from the individual ethical and moral values of each person. The economic crisis of 2008, as well as the environmental issue that for years has been at the centre of intense media coverage, has pushed this generation to develop a sustainable approach not only to investment, but also to consumption, lifestyles and mobility. This is the reason why many of their consumer choices are aroused by the desire to preserve the territories and the environment. Several studies (e.g. Carty, 2019; Cavagnaro *et al.*, 2018; McDonald, 2015; Ottman, 2011; Syngellakis *et al.*, 2018) highlighted that young travellers show a certain sensitivity in their choice of facilities, preferring, given the conditions of the same hidden costs, to adopt eco-friendly policies (such as the use of natural raw materials, recycling or energy saving).

Another eco-sustainable attitude adopted by young travellers around the world is to avoid printing paper tickets, but carry the digital version of their reservation or use QR codes on their technological devices. It dawns on us that sharing transportation and house sharing are not only solutions adopted to save money, but also strategies implemented to reduce air pollution. In China, for example, the electronic train ticket has already been implemented. Passengers no longer use paper tickets to get on a train. They can swipe their electronic ID card on the real-time communication app Wechat or on the third-party payment platform Alipay, or simply use their physical ID card before getting on the train. In big cities such as Beijing, Shanghai and Hangzhou, buses and subways also allow the use of electronic tickets in the form of a QR code.

On showing the QR code on the app or on Alipay at the station, the ticket fee is automatically deducted through the e-bank present on the app. This digital solution not only speeds up ticket purchasing and travel operations, but in times of a pandemic it has also become a tool for monitoring infections. In fact, starting from December 2021, local public transportation authorities in Hangzhou, the capital of East China's Zhejiang province, announced that the 'site QR code', which is embedded within Alipay, has been updated to track the major transportation hubs that people have visited. In other words, the code has also become useful in helping staff check the health status of people. In China, checking passengers' health QR codes and travel records for the previous 14 days has become an important part of epidemic prevention and control measures.

The practice of electronic ticketing is extending to London and many other cities around the world, where people can use their credit card instead of a ticket to pass the turnstiles.

This new approach will bring long-term benefits to urban environments: the use of fewer private cars will reduce traffic congestion, improve air quality in large cities and raise the living standard of their inhabitants. Fewer parking spaces will be needed and many urban areas can be recovered and dedicated to greenery. The environmental benefits of car sharing will be further enhanced by the use of electric vehicles, an important combination to encourage the spread of increasingly sustainable mobility.

Young people's attention to environmental issues is also evidenced by the growing number of sustainable investments. The data from Eurosif (2019), the organisation that promotes sustainable finance in European financial markets, are clear: Europe is recognised as an international leader with a weight of around 50% of the masses managed on the basis of SRI criteria worldwide. In this scenario, the surprises come precisely from millennials, who are approaching responsible investment more than senior investors did. Data empirically confirm that the new generations have a much more developed social and environmental sensitivity than older generations. Many people are increasingly interested in this type of investment, although sustainable solutions are very often not adequately favoured.

Tourism as a Life Experience

As highlighted in the previous chapters, a distinctive trait across the younger generations is their great interest in experiences, especially in the tourism sector. Over time, the holiday ends up taking on a special meaning in young people's biographies in which they can know themselves, the world and others, through direct contact with territories, inhabitants and material and immaterial cultural assets. In the tourism sector, however, getting in direct contact with the local population represents an opportunity for personal identity and cultural growth. This sociological interpretation of contemporary tourism allows to argue that collaborative consumption is, in effect, a different form of sociality, which, for this reason, is capable of producing social capital.

This awareness has led to the coinage of the term the 'experience economy' (Pine & Gilmore, 2001), in reference to the attitude of brands to attract younger consumers by trying to establish a deep dialogue with them. The willingness shown by new generations to obtain experiences has led companies to rethink their marketing strategies, offering goods and services that are not merely reaping benefits for customers but are also capable of stirring emotions among them. The most recent research in the marketing field has highlighted that from millennials onwards the behaviour of purchase is carried out first for an emotional reason and second for rational consumption value (e.g. Hunnicutt & Pine, 2020; Pine & Gilmore, 2019). In other words, young people no longer want to spend their money for accumulation, but rather for the enrichment of their personality. This change in attitude marks the transition from an external possession, typical of older generations, to an internal enrichment. It is no coincidence that the number of experience consumers has increased by 70% since 1987 in the US. The same growth is also recorded in Europe. These trends are confirmed by the Global Consumer Survey 2018 (Accenture, 2018), according to which the experience economy represents great potential for companies. Consumer spending on experience is expected to increase from $5.8 billion in 2016 to $8 trillion in 2030. China, as a huge and potential market, holds a special place in the Asian marketing scheme of luxury fashion brands. Upon stepping into China's market, Canada Goose, the Canadian brand famous for its down jacket, established a cold room in its Beijing-based store. It is a room, just as it is named, built of ice with a temperature of −27°F. Customers can test the heat retention property of Canada Goose's famous products in such a room. This cold room project not only satisfies the curiosity of its customers but also demonstrates its unique product function in the winter season. It is a marvellous example of experiential retail that serves the purpose of shortening the distance between young customers and the brand itself. By launching pop-up stores, offline salons or VIP tea parties, the brand gets the opportunity to make face-to-face contact with customers; on the other side, customers not only feel the product through touch and vision, but also feel an unprecedented intimacy that they have never felt before. In 2011, Prada began a new project with a focus on the Chinese market: restoring a historic garden villa in Shanghai. As the brand's official website describes,[1] the restored villa is 'an international cultural synthesis: a dialogue between Milan and Shanghai that pays tribute to the family who has originally commissioned this historic residence, to the Chinese architects and artists that created it and the team of Chinese and Italian craftsmen who were able to bring back its rightful splendour'. It was also used for the runway of Prada Resort 2018 and an exhibition regarding the restoration and the brand's other architectural demonstration. While appreciating the brand's latest season and the cultural and architectural history, customers bind themselves emotionally, without even noticing it, with the brand in the villa which combines the past with a modern style.

According to a study carried out by Euromonitor (Dutton, 2018), companies that also offer experiences with a high participatory value in the tourism sector stand a greater chance of acquiring potential customers, especially those belonging to the younger segments of the population. The study estimates that, looking only at the European area, purchases related to the experiential component will reach $8 trillion within a decade. The user's participation in the construction of the emotion can be more or less active. Thus, according to Pine and Gilmore (2019), there are four types of experiences: 'entertainment experiences' are those in which people passively experience the event. On the opposite side there is what the authors call 'educational experiences', during which the individual not only actively participates in the event, but their direct involvement also represents a learning moment, from which the participant comes out with the awareness of having learned something they didn't know before. In the middle, there are the 'aesthetic experiences', which occur when the individual immerses themselves in the event, but remains a passive user: this is the case, for example, of exhibitions. Finally, there are also the so-called 'escapist experiences', which occur when the individual is not only immersed in the experience, but participates in the initiatives that have been actively prepared for them, being involved, for example, through recreational activities.

In the near future, thanks to modern interactive technologies, it will be possible to solicit and capture the user's attention, allowing one to feel more involved and thus establish a closer relationship with the brand. This relationship between companies and consumers allows to create a relationship that goes beyond the mere commercial aspect. People who let themselves be involved by companies and seek experiences with them can be considered affiliates rather than just customers. Latest technologies make possible active interaction between the user and goods, services and territories: the internet of things (IoT), machine learning, artificial intelligence and big data analytics allow, today more than ever, to refine user engagement strategies.

Recent predictive studies (e.g. Gilovich *et al.*, 2014; Gilovich & Kumar, 2015) have emphasised that the experience economy will tend to grow further in the future for a purely psychological reason: taking part in events makes consumers feel happier than buying material goods, such as furnishings or clothing. Experiences are preferred because they do not end at the moment of purchase, but linger in the minds of people as important parts of their lives. Tourist experiences, in this sense, are gained by travellers as experiences capable of enriching their knowledge. Individuals who actively interact with local communities and with other people they meet during their trip experience emotions that not only give immediate benefits, but also remain alive in their memory, making them feel better even when the trip is over.

More specifically, regarding the tourism sector, the platforms mostly used by young people from all over the world that are located within these spaces are those that allow the rental of houses or rooms. Among the most used services, Airbnb stands out as a portal that connects the demand and supply of short-term accommodation and perfectly embodies the aforementioned philosophy of life. The site allows a home exchange between both sides of demand and supply or a whole or partial apartment rental to tourists. Airbnb has based all its businesses on the search for authenticity. Users enjoy this service not only for economic reasons (to save money), but also to have direct contact, as a peer, with local residents. Other platforms that embrace the same principles are BeWelcome and CouchSurfing, which allow people to access the home (or a portion of it) of another member and stay for free for a limited period of time. Other services that meet the interest of young tourists are those that allow them to dine at the home of people who live in the place they intend to visit, so as to taste the local cuisine and, at the same time, enjoy the company of locals who can share their knowledge of the territory and show their lifestyle and part of their everyday life.

For some tourists, the direct involvement of the local residents is very attractive because it allows travellers to know the places where they stay through the eyes of those who live there. Moreover, according to Monaco (2019b), this type of practice also activates a tourist experience that could be defined as 'transfigured': people who receive guests from other parts of the world, host in their own home new ways of living, thinking and behaving different from their own, enjoying all the benefits sought after in post-modern tourist experiences, even without moving. As described by Trua (2016) about her experience as a host on the CouchSurfing.com portal, having people from other countries in her home transformed her days into a continuous exchange of information, anecdotes and experiences. As is evident, transfigured tourism is capable of producing a new, hybrid spatiality, which is not only a place of origin, but also a destination. These types of practices are able to activate a process whereby people become tourists in their own home. The people who host open their doors to people who come from other places, who enter into their daily lives and are able to provide elements capable of significantly affecting their identity. This tourism accommodation method, in fact, supports reciprocal activities at the basis of which the authenticity of the relationship can be found, together with sustainability in consumption and processes of cultural contamination.

The Shadows of Sharing

The interest matured over time towards collaborative consumption and the use of platforms that are above all allowing young people to put into practice new forms of tourism experiences are counterbalanced by a series of factors that in some way negatively affect the sharing economy practices. Firstly, it is clear that it is not possible not to refer to the difficulties brought about by the coronavirus emergency, which somehow generated, albeit indirectly, a sort of mistrust between people who do not know each other, lowering the levels of trust because of the possibility of contagion (Monaco, 2021). Clearly, as mentioned earlier, various areas and services which fall within the sharing sector have been affected in different ways.

As for mobility, some of the ride-hailing giants have faced a major crisis globally. In 2020, compared to the previous year, Uber recorded 80% fewer runs and looked for financial protection, thus laying off a quarter of its staff. To prevent bankruptcy, in late March the company Bird was forced to lay off over 400 employees, a third of its workforce. In the post-Covid period, however, the situation is gradually recovering, also because when people have a choice, they prefer to use these kinds of services instead of taking public transport. Therefore, in this scenario, shared mobility solutions and carpooling should follow the regulatory provisions and the hygienic/sanitary precautions necessary to guarantee the absolute safety of travellers. More precisely, the fear of contagion that still lingers among people could represent the impetus to register margins for growth, provided that the companies engaged in mobility manage to spread in the collective imagination the idea of being attentive to the well-being and safety of their passengers. They must guarantee not only adequate sanitation but also above all constant and careful monitoring of their crews and the condition of their cars. In China, where the coronavirus initially broke out, Didi (an online car-hailing platform like Uber) also experienced a downturn in its passenger flow volume, revealed by Liu Qing, the CEO in an interview with CNBC. In response to the outbreak of coronavirus in early January 2020, Didi quickly adopted necessary and urgent measures: flu masks and disinfectants were dispensed to every Didi driver; a transparent cover was put in place to separate the passenger and the driver; every driver was to record his or her body temperature on an app during certain hours of the day; the driver and passenger were to follow the requirement of wearing a mask, otherwise the driver had the right to refuse the passenger, etc. In May 2020, the passenger flow volume had recovered 60–70% of that before the outbreak of the coronavirus, five times higher than its lowest volume back in February.

As expected, globally the hospitality sector experienced a major setback during the lockdown. Airbnb lost around 90% of its bookings and laid off a quarter of its employees, because it had to repay users who had or wanted to give up their vacation, which accounted for orders of over a billion dollars. Undoubtedly, this whole situation led to the collapse of its market value. To deal with the emergency, in the post-pandemic period, the giant implemented a series of strategies: it encouraged its hosts to apply flexible cancellation policies, implemented protocols for cleaning and sanitising environments that all members must necessarily follow and activated some promotional campaigns to guarantee more advantageous prices to travellers who decide to stay in homes for a longer time. In addition, to react to the predicament, Airbnb has tried to make the most of the potential offered by the network, offering its users the opportunity to enjoy online tourist experiences. Even before the resumption of international mobility, the platform has somehow tried to give an alternative orientation to its business, by inserting on its portal a specific section dedicated to digital experiences. In this way, travellers, while staying at home, have the opportunity to enjoy the culture of the places they intend to visit. By paying a price varying from 1 to 40 euros, tourists from all over the world can take part not only in a series of digital experiences such as language courses, cooking, guided tours, visits to museums and monasteries, and dance sessions, but also typical activities, such as shows, fairs and events. All activities are accessible through Zoom. In the first period, Airbnb offered its users the opportunity to enjoy over 100 digital activities spread across 30 countries around the world. The objective in the short to long term has been to extend the geographical perimeter within which to enjoy these experiences, involving more than 100 countries, in an attempt to allow people who are still held back by fear of contagion a wide range of places to discover and visit on demand.

Other elements that negatively impact on the sharing of experiences are represented by the possible low quality of services, by the lack of trust in people known only online and by scepticism in sharing something personal. These aspects are accentuated by the absence of a single and definitive regulation that protects travellers and hosts.

As reported by the European Parliament (2015), the rapid development of the sharing economy is not only an opportunity, but also a challenge. Indeed, much of the sharing economy, although its share is still very limited, is affecting the tourism sector. Its creative capacity and the advent of an alternative mode of competition are the principal opportunities. However, taxation and legislation, which are not always straightforward and well defined, are the main challenges.

A qualitative study (Ranzini et al., 2015) realised in six European countries (Germany, Italy, the Netherlands, Norway, Switzerland and the UK) selected in order to achieve a geographical spread of the sharing economy across Europe revealed that millennials reported occasional experiences of discrimination based on race, gender or sexual orientation. For many of the interviewees, the fear of running into this type of negative experience represents a brake on which they think it is necessary to legislate.

Under these circumstances, if these forms of relational tourism reacquire a greater centrality in future mobility experiences, an intervention by the legislator that guarantees a transparent and disciplined management of the new collaborative services appears necessary.

Conclusion

To summarise, it is possible to argue that members of new generations adopt a more holistic approach to happiness than what is manifested by members of previous generations. In this sense, they are supported by media and social networks which are blurring the line between the virtual world and reality. The attitude taken by young people allows to definitively deny the idea, long-standing also in the academic field, according to which one of the effects of new technologies would be the devaluation of the essence of the world (Mandich, 1996). New technologies and virtual spaces are important resources that contribute significantly both to enriching the world view and to widening the range of possibilities available to them. In other words, new technologies make possible a confluence between absence and presence, between local and global, between proximity and distance, increasing and multiplying the opportunities for personal growth, social interaction and cultural exchange (e.g. Colombo, 2005; Couldry & McCarthy, 2003; Nyìri, 2005; Monaco, 2018a; Moores, 2012).

From this critical angle, it can argued that the experience economy can be interpreted both as a concrete response to less economic availability and as a strategy put in place to enjoy a state of lasting happiness which stands the test of time. From millennials onwards, young people began to gain consumer experiences as a continuous search for authenticity, for truth, which can be achieved through a peer-to-peer dialogue with brands. This situation is urging brand companies to adopt unprecedented interaction strategies which in the future may prove successful in capturing the attention of younger generations. In particular, in the tourism field, companies must be able to communicate skilfully on the strength of social networks, activating equal conversations and offering experiences that are considered interesting, not merely the goods and services that they intend to sell, even better if undertaken in a sharing context. Among the companies in the vanguard, for example is the airline British Airways, which is investing heavily in the related project of customer experience through multisensory marketing proposals. Among the various initiatives carried out in recent years, the smell of grass pervading the lounge of terminals made the passengers feel more relaxed while waiting for their plane, or the game 'Top 18 2018' that involved people who were visiting Covent Garden in London, rewarding the winners with tickets to 1 of the 18 top destinations served by the airline.

In the future, companies will have to give ever greater appropriate prominence and visibility to their social commitment and the ethical codes underlying their decisions. It is no coincidence, for example, that Airbnb, despite the economic difficulties encountered during the climax of the emergency of Covid-19, has launched a series of laudable initiatives, aiming at demonstrating to its user base that it also cares about social welfare, as well as the economic profit. In the specific case, during the period of maximum emergency on a global level, the platform invited its hosts to open the doors of their homes to offer free or advantageous accommodation to medical and healthcare personnel. Simultaneously, Airbnb has activated a fundraiser to offer stays to professionals who are personally involved in the health emergency. It has been a strategy that has managed to promote home sharing, but in a more responsible and supportive way, even in times of crisis.

In conclusion, emerging developments, regardless of the particular characteristics of the territorial background and the desires of the individual targets, may be an opportunity to reformulate those business models geared towards sustainable innovation, which may somehow deviate from the logic of benefit. Clearly, this perspective is not applicable to all the players involved in the tourism sector.

We must not forget that in the past the sharing economy has also caused negative impacts on tourist destinations (e.g. Bocken *et al.*, 2014; Jackson, 2009). Not all sharing economy models have been necessarily 'green' or 'fair', since some of them followed a basic economic rationale. Besides, some sharing economy models were at risk of harming the rights of workers, since models replaced existing stable jobs with unstable, poorly paid and sometimes even exploitive or illegal work relations.

As reported by Verboven and Vanherck (2016: 24): 'Sharing economy business models do not always maximize societal benefits due to negative externalities, such as the wrong allocation of subsidies, an increased income disparity and the lack of systematic health and safety controls'.

Looking at the medium to long term, despite the wide range of subjects involved in a transversal way, the future has to be oriented towards a responsible social economy, based on innovation, and sustainable development, in which the central role of technology has to be evident as an indispensable means to improve the protection of rights and social justice.

With the emergent paradigm that keeps changing, both the tourism sector and the new generations have to take into consideration common goods, respect for workers' and citizens' rights and their active involvement. Only in this way will it be possible to reach a 'just transition', addressing social and environmental justice concerns within and across nations (Morena *et al.*, 2020).

The first step to driving forward the just transition is to avoid repeating the mistakes of the past and, above all, to listen to the most innovative and responsible voices coming from young generations, thus also eradicating social asymmetries and inequalities.

Note

(1) https://www.pradagroup.com/

Part 3
Generations, Gender and LGBT Issues in Tourism

5 Gender, Generations, Tourism

Elisabetta Ruspini

Introduction

This chapter examines the relationship between generational change, gender and tourism, specifically focusing on the younger generations (millennials and members of Generation Z). The chapter aims to integrate a gender perspective into the generational analysis of tourism experiences and activities. The gender perspective essentially conceptualises gender as a social structure that organises society into different and unequal categories based on physical and physiological differences between males and females. The concept of gender expresses the institutionalised system of values, norms and social practices aimed at constituting people as two significantly different categories, organising social life and relations on the basis of that difference and determining the functions and responsibilities of each sex (Lorber, 1994; Millett, 1971; Ridgeway & Smith-Lovin, 1999; Risman, 2004; Rubin, 1975; Scott, 1986). The social construction of gender differs over time and by specific sociocultural contexts and is produced, sustained and renewed through an intricate arrangement of practices and shared understandings within a given society. 'Doing gender' is as an ongoing activity embedded in everyday interaction (Thompson & Armato, 2012; West & Zimmerman, 1987: 130).

As written (Swain, 2005: 247), tourism, as leisured travel, is built of human relations, and thus impacts and is impacted by global and local gender systems (Kinnaird *et al.*, 1994). Tourism experiences are grounded in, and influenced by, our collective understanding of the social construction of gender (Hall *et al.*, 2003). Gender is a key analytical and explanatory variable in the process of social construction of any tourist activity. Tourism research has shown growing interest in gender-sensitive research, demonstrating that this area of study has a wide margin for improvement (Chambers *et al.*, 2017; Pritchard, 2018; Segovia-Pérez *et al.*, 2019). Outlining the trajectory of tourism gender research, Cohen and Cohen (2019) point to the emergence of research on tourism and gender in the 1980s, and highlight as foundational the special issue on 'Gender in Tourism' in *Annals of Tourism Research* (Swain, 1995). As highlighted by Figueroa-Domecq and Segovia-Pérez (2020), a gender perspective was included in different areas within the field of tourism studies: consumer behaviour research, tourism activities and behaviour tourism development, relationships between hosts and tourism, unequal gendered power relations in tourism, and gender stereotypes (Kinnaird & Hall, 2002; Swain, 1995; Pritchard & Morgan, 2000). Recently, a book containing tourism research created by women has been published (Correia & Dolnicar, 2021).

Notwithstanding these significant contributions, tourism enquiry has been reluctant to engage gender-sensitive frameworks. Several scholars (Figueroa *et al.*, 2015; Figueroa-Domecq & Segovia-Pérez, 2020; Pritchard, 2018; Pritchard & Morgan, 2000; Westwood *et al.*, 2000) underline the existing lack of critical thinking and critical gender research, resulting in a relative scarcity of tourism studies that fully incorporate gender in the evaluation of the tourist experience and that apply a feminist perspective. For example, there is a growing concern about the production of knowledge in tourism and the situation of women within tourism academia (Munar, 2017; Pritchard & Morgan, 2017; Tribe, 2006). It thus seems necessary to broaden and deepen feminist and gender research in tourism studies.

Due to gender ideologies and stereotypes, women and men participate in and experience tourism differently as both consumers and producers (e.g. Gilli & Ruspini, 2014; Pritchard *et al.*, 2007; Swain, 2005; Swain & Henshall Momsen, 2002; UNWTO, 2011b). Throughout history, traditional views of gender that emphasise the value of and the need for distinctive roles for women and men have severely limited women's mobility (Wearing, 1998; Wearing & Wearing, 1988). Mobility constraints are closely related to gender ideologies that connote domestic space/work as women's space/work (Kelly, 2012). Prior to World War II, participation rates in leisure activities outside the home were heavily differentiated with respect to gender, with most female activities being indoors (Collins & Tisdell, 2002). Due to a number of factors that will be discussed in the next paragraphs, the mobility of women has progressively changed during the past several decades and today the number of women travellers is increasing globally. The current historical phase is however marked by a strong tension between past and present. On the one hand, traditional gender ideologies, infused with notions of space, place and mobility (Hanson, 2010), are still powerful in contemporary society and gender gaps and dichotomies still exist in leisure activities (Godtman-Kling *et al.*, 2020; Jackson & Henderson, 1995; Khan, 2011; McGinnis *et al.*, 2003; Shaw, 1994; Wearing, 1998). On the other hand, social change has favoured a drawing closer of male and female life courses from different points of views. The best known of such changes include an increase in women's education and employment, delayed entry into adult life, a shared and lesser inclination for marriage and procreation, the assumption by women of responsibilities which previously belonged exclusively to men (Beck & Beck-Gernsheim, 2001; Hantrais, 2004; Jacobsen *et al.*, 2015; Lamanna & Riedmann, 2009; Oppenheim Mason & Jensen, 1995). Processes of change are exerting a major influence on the travel motivations and behaviour of the new generations, challenging contemporary tourism. Today it seems that young women are travelling more than young men (Tilley & Houston, 2016) and female solo travel is on the rise globally as women are increasingly choosing to travel alone.

In this context of increasing women's participation in leisure and tourism, the global pandemic and its unprecedented challenges may reverse positive trends. The outbreak of the coronavirus disease (Covid-2019) and the resulting restrictions imposed to fight the rapid spread of the virus have dramatically affected life courses, public services, economies and opportunities worldwide. With almost 240 million confirmed cases of Covid-19 globally, including more than 4.8 million deaths reported to the World Health Organisation (WHO Coronavirus Disease Dashboard, 15 October 2021), the Covid-19 pandemic has been defined as the third and greatest economic, financial and social shock of the 21st Century, after 9/11 and the global financial crisis of 2008 (OECD, 2020). The drastic social distancing measures that have become the primary policy prescription for combating the Covid-19 pandemic have increased isolation and caused widespread concern and fear. Due to full or partial lockdowns (social and self-isolation rules, global mobility and travel restrictions, border closures, closure of tourist attractions), travel and tourism are among the most affected sectors by the Covid-19 crisis (UN Women, 2020a; UNWTO, 2020a, 2020b, 2020c). There is also a growing concern that the Covid-19 crisis will disproportionately affect women and girls because of existing gender inequalities that can be exacerbated in the context of health emergencies.

Based on these complex premises, the chapter first offers a brief overview of the history of women's travel and constraints on women's access to leisure based on an overview of theoretical as well as empirical advances in research. It then analyses some of the peculiarities of the millennial and Gen Z travellers, addressing the transition of gender roles and values. It lastly reflects on the impact of the pandemic on women's travel and on women's tourism in a post-Covid-19 world, also revealing that both generations remain resilient and seem determined to help drive positive change.

The Past

Access to recreation is not (and has not been) equally distributed across society. Women have long been the disadvantaged gender in leisure, travel and tourism (Khan, 2011). Previous studies have helped in understanding the reasons behind the constraints to leisure for women

(Henderson, 1991; Jackson & Henderson, 1995; Wearing, 1998). Sociodemographic characteristics play a key role in the prevalence and extent of this access (Godtman-Kling *et al.*, 2020), including family situations, ethnic groups, social class and income level. Constraints to recreation are more pronounced for people who are in non-dominant groups and gender is an important determinant (Shores *et al.*, 2007). The system of two distinct genders with opposing traits has been, and is, powerful in contemporary society, including within leisure and tourism (McGinnis *et al.*, 2003). Traditional gender ideologies emphasise the value of distinctive roles for women and men, suggesting that men should fulfil their family roles through breadwinning activities while women should be caretakers. These ideologies equate, on the one hand, women and femininity with the home, the private, with domestic spaces and restricted movement (which translates into familiar and routine interactions) and, on the other hand, equate men and masculinity with the public, with urban spaces and wide-ranging movement (which translates into interactions that bring excitement, challenges, new experiences, encounters with the unknown: Hanson, 2010: 9). Women have been assigned the role of primary caregiver because they can give birth and, through the socialisation process, caregiving is understood by women as both normal and inevitable, as a source of fulfilment of their natural role, and as a moral obligation.[1] Indeed, they feel societal pressure when they decide against offering their caregiving services (Friedemann & Buckwalter, 2014). Due to traditional patriarchal ideologies, men and women have unequal access to leisure activities and the time to pursue them. Because of their caretaking role within the family, involving unpaid domestic labour often in addition to formal employment, women tend to face stronger leisure constraints, while men spend more time and have access to a wider range of leisure opportunities (Hargreaves, 1989; Shaw, 2001; Wearing, 1990, 1998). Often the main constraints involve a lack of resources such as time, independent income, childcare and safe transport, and the most severe limitations are experienced by specific groups of women (Deem, 1982; Green *et al.*, 1990) including unemployed women, single mothers, women with young dependent children, women with substantial caregiving responsibilities, migrant women and elderly women. Gender differences in mobility have been well documented, with women travelling shorter distances compared to men (Pooley *et al.*, 2005; Tilley & Houston, 2016). And even if women have been travelling for centuries they have, until recently, generally been overlooked in the travel, tourism and exploration literature (Wilson & Little, 2005: 156; also Clarke, 1988; McEwan, 2000; Towner, 1994).

In the past, it was not socially acceptable for women to travel (Khoo-Lattimore & Wilson, 2017) and certainly not on their own unless accompanied by a male family member. According to Robinson (1990, cited in Richter, 1994: 392) 'until the sixteenth century, to be a woman, travel and remain respectable one had to be generally either a queen or a pilgrim'. The social structure of 16th-century Europe allowed women very limited opportunities outside their households. The proper place for women was in the home and public leisure was a male domain: only privileged women could fail to conform to this norm (Khan, 2011; Rybcyzynski, 1991). The 17th and 18th centuries were not an era of drastic changes in the status or conditions of women. Women continued to play a significant, though not acknowledged, role in society, primarily through domestic activities. Education was seen as a way of making women better wives and mothers but young girls did not need to go to school to learn: they were educated at home by their mothers who acted as role models. Even in highly literate groups, girls were often taught only basic reading and writing, as well feminine activities such as needlework and dancing (Friedman, 1985). In the 18th century, securing a good marriage was the most important, if not the only goal of girls, especially if they belonged to the upper class which required that women learned all the tasks considered necessary to make a good wife and mother (Jordan, 1991). Travel was irrelevant to the education of women, unless it functioned to support family goals, was justified for religious pilgrimages or for health and medical reasons. In contrast, men belonging to privileged social classes were given the opportunity to undertake educational travel. As such, the unchaperoned travel of single women compromised marriage prospects by diminishing a woman's reputation (Richter, 1994; Rybcyzynski, 1991).

Beginning in the mid-17th century, it became the custom of the upper class to send their sons abroad in their early twenties: we are referring to the Grand Tour (17th century to mid-19th century) which was considered to be the best way to complete a gentleman's education (Cohen, 2001). Although generally associated with the English aristocracy, wealthy families from all over Europe (such as Denmark, France, Germany, the Netherlands, Poland and Sweden) sent their sons away to expand their cultural experience. This extended tour typically involved two to four years of travel around continental Europe and included an extensive sojourn aimed at introducing young men to the art and culture of Italy, seen as the cradle of Western civilisation. Young men were often accompanied by older and more experienced tutors, who might also supervise studies. The Grand Tour began as an almost wholly male rite of passage. However, by the latter part of the 18th century, rising incomes, cheaper and more reliable travel, and changes to the political landscape of Europe made travel increasingly accessible to women (Gleadhill, 2017), provided they were in the company of male relatives (Dolan, 2001; Lindeman, 2017). Well-known examples of female grand tourists are Lady Mary Wortley Montagu (1689–1762), Lady Anna Riggs Miller (1741–1781), Hester Lynch Piozzi (1741–1821), Mariana Starke (1761–1838), Lady Elizabeth Holland (1771–1845) and Catherine Wilmot (1773–1824) (Olcelli, 2015; Watts, 2008).

Towards the Present

Nineteenth-century travel became more democratised and feminised (Haynie, 2014). If many of the early female travel writers were often women belonging to religious orders, aristocrats or diplomats wives, the 19th century saw an increase in the number of women travelling independently and alone, no longer travelling merely as accompanists to their husbands (Robinson, 1994). By the mid 1800s, the growing numbers of independent women travellers became a notable trend (Pemble, 1987). This development was built upon a number of changing social, cultural and economic factors: advances in educational opportunities for women and girls, industrialisation and urbanisation, the creation of a middle class that enjoyed the benefits of the new prosperity, scientific innovations and technological improvements. In the 19th century, obligatory primary education for girls became more widespread and women began to play central roles in education – as teachers and as learners – even if scholastic institutions created different curricula for boys and girls and, in the Western world, universities did not begin to accept women as students until the second half of the century. With the development of modern means of transportation – rail travel, in particular, eased some of the challenges of travelling – women increasingly took part in travel (Bourguinat, 2016).

At this point in history, the first feminist movements formed in Europe: from the 1850s in England and Scandinavia, then in Western and Central Europe, and then during the 1870s and 1880s this first feminist wave extended to Eastern and Southern Europe (Briatte, 2016). Women began to redefine traditional concepts of space and boundaries and to defy societal and gendered conventions by travelling, often solo, into unexplored environments (Sambuco, 2015) and the number of female travellers grew in importance and numbers. Well-known examples are Ida Pfeiffer, an Austrian explorer and travel writer who undertook two world tours (1846–1848 and 1851–1855); the Swedish feminist and writer Fredrika Bremer, who travelled extensively through the US between 1849 and 1851; the British travel writer Helen Lowe, who travelled to Norway (1856), Sicily and Calabria (1857) with her mother; Constance Frederica Gordon-Cumming, a Scottish travel writer and painter who travelled extensively in Asia and the Pacific; Mary Henrietta Kingsley, a British traveller and ethnographer of West Africa (1893–1895); the British writer and archaeologist Gertrude Bell, one of the few women who travelled to the Middle East in the 19th century; and Isabelle Eberhardt, a Swiss explorer who lived and travelled extensively in North Africa, often dressed as a man.

The development of tourism in the 20th century can be divided into different periods. A first period of rapid growth, the stagnation as a result of World War I, remarkable growth in travel and tourism in the interwar period, World War II, and the exponential increase of tourism in the post-war period. During the latter half of the 1930s, the seeds of mass tourism were planted. Following pressure from trade unions, European workers were typically granted an average of one to two weeks paid vacation (Vukonić, 2012). In 1936, the International Labour Organisation (ILO) adopted the 'Holidays with Pay Convention (no. 52)', which provided for an annual holiday with pay of at least six working days after one year of continuous service (12 days for workers under 16 years of age). Paid annual holidays became widespread after World War II. Beginning in the 1930s, the growing availability of the motor car stimulated tourism further and automobile tourism grew rapidly. During the interwar years, aircraft began to play a role in tourism (Sezgin & Yolal, 2012). Economic and social transformations following the end of World War II democratised tourism in Western countries, making holidays accessible to all social classes and not just an activity for the wealthy élite (Gardiner *et al.*, 2013: 311). Cultural, economic and technological developments that changed tourism in this historical period – and influenced the formative experiences of the younger generations, and thus its members' beliefs, values, and behaviour – are linked to the post-war economic boom and have been summarised as follows (Sezgin & Yolal, 2012): improved economic prosperity; the growing sense of democracy and equality; the increasing availability of private transport; the rapid growth of air transportation; the television explosion that became a cultural force in the 1960s; the prominent function of the tour operators; and the rising number of people enjoying paid vacations. By 1960, most employees in continental Europe were legally entitled to two weeks of paid holiday (three in Norway, Sweden, Denmark and France), and increasingly they took that holiday away from home (Judt, 2005: 340). The counterculture movements from the early 1960s through the 1970s profoundly affected generations around the world, introducing new mindsets and ways of travel. The hippie movement had a deep impact on youth values, with new behaviours and sensory experiences promoting the desire and feeling of freedom (Pereira & Silva, 2018). Initial domestic demand was followed by the rapid growth of international tourism in the late 1960s and early 1970s. At the same time, consumers began to seek more individualised experiences than in the past and, as a result, tourist product offerings became more varied and commoditised (White, 2005). In response to the economic situation and strategic innovations, commercial tour operators and travel companies introduced new destinations and vacation types (Gardiner *et al.*, 2014; Gyr, 2010). Holidays and travel became accessible to ever broader strata of the population, such as social groups defined by age and gender (women, single persons, elderly people), that took advantage of specific products tailored to their various demands.

The changing travel patterns of women must be seen in light of the complex interactions of major sociocultural and economic trends: the cultural revolution of the late 1960s and 1970s, the influence of the post-World War II feminism movement (second-wave feminism), equal opportunities legislation, the growing education of women and their increasing labour force participation, the expansion of the service sector and the knowledge economy, and greater access to mobility resources (Khan, 2011). As noted by Rosenbloom (2006), in the past three decades most industrialised countries have seen three main related trends. The increasing involvement of women, particularly those with children, in the paid labour force; changes in the distribution of household responsibilities; substantial alterations in household and family structure (the decrease in the number of children, the increase in one-parent families and the growth of single-parent family households, combined with the aging of society). These trends have triggered significant travel-related changes, boosted women's travel and reduced gender differences in mobility (Frändberg & Vilhelmson, 2011; Mcquaid & Chen, 2012; Rosenbloom, 2006). The increasing number of women working outside the home and pursuing careers had led to an equal need by men and women for a holiday to recoup from the demands of work (White, 2005). While men were (and still are) more likely to have a driving license compared to women, the proportion of women having a license has been increasing at a faster rate, and travel companies dedicated solely to women have increased rapidly too (Tilley & Houston, 2016). Women's travel has become a major phenomenon over the past few years, including business travel, adventure travel and ecotourism. One of the most remarkable aspects of this trend is that women belonging to different generations are increasingly choosing to travel alone (Pereira & Silva, 2018; Wilson & Little, 2008). This trend has been investigated by a number of researchers (Bialeschki, 2005; Bond, 1997; Chiang & Jogaratnam, 2006; Matthews-Sawyer *et al.*, 2002; McNamara & Prideaux, 2010). Existing studies have identified the motivations of solo independent women travellers. From the desire to challenge oneself (Jordan & Gibson, 2005; Wilson & Little, 2005), to the wish to develop a feeling of autonomy and independence (Butler, 1995; Wilson & Harris, 2006; Wilson & Little, 2008), to the search for new experiences, relax, self-discovery and self-awareness (Bond, 1997; Chiang & Jagaratnam, 2006). Another relevant reason is the importance of socialisation, the interaction with other travellers and the native people (Jordan & Gibson, 2005; Wilson & Harris, 2006; Wilson & Little, 2005). Solo women travellers today represent a growing and influential market segment.

However, gender differences in mobility are still strong due to heavy inequalities in employment (horizontal and vertical segregation), care and domestic work, traditional gender ideologies and gender stereotypes (Transportation Research Board, 2006). As noted by previous research (Whyte & Shaw, 1994; Wilson & Little, 2008: 181–182), women solo travellers are challenged by security concerns and have to negotiate the social stigma concerning appropriate travel behaviour. A 'geography of women's travel fear' (Valentine, 1989) is yet evident in the solo female travel experience. This concept suggests that women's travel and motivations for independent tourism experiences are still governed to some degree by a structure of patriarchal social control. Tourism is a sexualised and gendered space and, within it, solo female travel is considered as a voluntary risk-taking behaviour. Solo women travellers experience fears that are based predominantly on an anticipation of male violence and harassment (unwanted male gaze, inappropriate comments, physical and/or sexual assault) when travelling.

Between Present and Future: Millennials and Gen Zers

As written in Chapter 1, generation, in the Mannheimian sense, is a concept that can be used to identify people whose life courses are forged through the same conditions (Mannheim, 1928, 1952). Each generation has specific characteristics because cultural, technological, economic and social events differently interact with its members' formative experience and life courses. This interaction affects a generation's collective identity. Previous research suggests that the historic trend towards egalitarian views on gender roles can be attributed to generational change. Educational expansion (Thijs *et al.*, 2019) and the rise of women's labour force participation are the most frequently mentioned explanations for the increase in support for gender egalitarianism (Brewster & Padavic, 2000; Brooks & Bolzendahl, 2004; Cotter *et al.*, 2011). On the one hand, it is argued that education has a liberalising influence, transmitting ideas about diversity and equality, countering gender stereotypes and increasing individuals' openness to new perspectives on the roles of women and men in the public and private spheres (Bolzendahl & Myers, 2004; Thijs *et al.*, 2019; Vogt, 1997). The higher the level of education in society to which people are exposed during emerging adulthood, the stronger people will support gender egalitarianism, and this effect will be stronger for women than for men (Thijs *et al.*, 2019). On the other hand, women's rising labour force participation has provided women with broader economic opportunities, increasing their access to earnings and occupation-based networks and altering patterns of inequality between and within genders (Morris & Western, 1999).

Millennials and Gen Zers are much better educated than their grandparents. For example, the report from the Pew Research Center 'Millennial Life: How Young Adulthood Today Compares with Prior Generations' (Bialik & Fry, 2019) clearly show that US millennials have higher levels of post-secondary education than earlier generations. Among American millennials, around 4 in 10 (39%) of those ages 25–37 have a bachelor's degree or higher, compared with 15% of the Silent Generation, roughly a quarter of baby boomers and about 3 in 10 Gen Xers (29%) when they were the same age. Gains in educational attainment have been especially steep for young women. Millennial women are about four times (43%) as likely as their Silent predecessors to have completed as much education at the same age (Bialik & Fry, 2019). As for Gen Zers, data from the Pew Research Center report 'On the Cusp of Adulthood and Facing an Uncertain Future: What We Know About Gen Z So Far' (Parker & Igielnik, 2020) suggest that older members of American Generation Z are on a different educational trajectory than the generations that came before them. They are less likely to drop out of high school and more likely to be enrolled in college. Among 18- to 21-year-olds no longer in high school in 2018, 57% were enrolled in a two-year or four-year college. This compares with 52% among millennials in 2003 and 43% among members of Gen X in 1987 (Parker & Igielnik, 2020). As to women's involvement in the labour market, the already cited Pew Research Center analysis (Bialik & Fry, 2019) shows that in 1966, when Silent Generation women were between 22 and 37 years old, a majority (58%) were not working while 40% were employed. For millennial women today, 72% are employed while just a quarter are not in the labour force. Boomer women were the turning point: as early as 1985, more young boomer women were employed (66%) than were not in the labour force (28%). Moreover, a 2015 PwC survey on 8756 female millennials (born between 1980 and 1995) from 75 countries, aimed at understanding how they feel about the world of work and their career, discovered that millennial women are more confident and career ambitious than the generations preceding them (PwC, 2015). A relevant number of millennial women (49%) starting their careers believe they can reach the very top levels with their current employer but almost half say employers are too male biased when it comes to internal promotions. Of the millennial women who are in a relationship, 86% are part of a dual career couple, with 42% earning equal salaries to their partner or spouse. Almost a quarter (24%) are the primary earner in their relationship.

In addition to this, both millennials and Gen Zers have become socialised into an increasingly egalitarian societal context, instilling stronger support for gender egalitarianism in these generations. As already mentioned, they have grown up in both more plural family settings (divorced families, blended families, single-parent families, living apart together [LAT] relationships, lesbian, gay, bisexual and transgender [LGBT] families, etc.), and in families with two working parents, where the mother and the father have started to share care responsibilities, and got the message that gender equality is possible and has positive impacts. Millennials are the children of baby boomers (and early Gen Xers), viewed as a strategic generation that reshaped life courses, lifestyles and institutions (Edmunds & Turner, 2005; Leach *et al.*, 2013). Boomers fought against race and gender inequality, participated in anti-war protests and supported sexual freedom. The cultural revolution of the late 1960s and 1970s – which was itself fuelled by a post-war prosperity that allowed people to give greater attention to non-material concerns and post-material values (Inglehart, 1977) – played a key role in reconfiguring views of gender relations, marriage and family life. Feminism as a political movement rose accordingly with economic growth, as a political challenge to the androcentric bias in this growth (Lang *et al.*, 2013). 'Second-wave feminism', which refers to the activism of the 1960s and 1970s, encompassed some of the most widely known feminist causes: equality in education and employment, abortion and birth control access, raising consciousness about gender-based violence (GBV) and marital rape, sexual liberation of women, and included campaigns in support of peace and disarmament (see, e.g. Ryan, 1992; Schulz, 2008). Second-wave feminism grew in post-war Western societies, among the student protests, the anti-Vietnam war movement, the lesbian and gay movements and, in the US, the civil rights and black power movements (Kroløkke & Sørenson, 2005). Starting from this period, traditional gender roles have undergone a major shift in most parts of the developed world. Millennials is, indeed, a generation regarded as being open-minded and more supportive of gender equality, LGBT rights and equal rights for minorities (see, e.g. Taylor & Keeter, 2010; Rainer & Rainer, 2011; Risman, 2018; also Chapter 6) and by changing attitudes towards marriage and family (Parker *et al.*, 2019). In this regard, in her study based on in-depth interviews with a non-representative sample of more than 100 young people, including the experiences of transgender and gender queer youth, Risman (2018) notes that millennials are pushing boundaries by not only rejecting traditional distinctions between the sexes, both at home and at work, but also refusing to accept gender categories altogether. However, as Risman (2018) explains, millennials vary greatly in how they understand and position themselves in reference to the gender structure as well as in the strategies used to negotiate the ongoing gender revolution. Millennials are, at the same time, getting married later and increasingly refraining from long-term commitments with respect to partnerships and child-bearing: both men and women want to first gain valuable life experience and establish themselves in the labour market before founding a family (Oláh *et al.*, 2018).

Generation Z's views resemble those of millennials in many areas (Parker & Igielnik, 2020) but Generation Z is moving toward adulthood with a liberal set of attitudes and an even higher level of openness to emerging social trends (Parker *et al.*, 2019). Among Gen Zers, traditional views of gender roles (women staying at home, men going to work) are no longer the norm in many countries. Members of Generation Z are more likely to have a positive view of interracial and same-sex marriage than their older counterparts, to identify as solely heterosexual and have greater contact with people who do not identify as just one gender (Deloitte, 2018; Ipsos Mori, 2018). Pew Research Center analyses[2] (Parker *et al.*, 2019) show that Gen Zers are more likely than millennials to say they know someone who prefers that others use gender-neutral pronouns to refer to them: 35% compared with a quarter of millennials. Among each older generation, the share saying this drops: 16% of Gen Xers, 12% of boomers and 7% of members of the Silent Generation say the same thing. The youngest generation is also the most likely to say forms or online profiles that ask about a person's gender should include options other than 'man' or 'woman'. Roughly, 6 in 10 Gen Zers (59%) hold this view, compared with half of millennials and 4 in 10 or fewer Gen Xers, boomers and Silent Generation members (see Chapter 6 for details).

The younger generations represent a break with hegemonic masculinity, as well. The number of men willing to question the stereotyped model of masculinity is growing and millennial men and fathers are more involved and egalitarian than men/fathers of previous generations (Ruspini, 2019; Ruspini *et al.*, 2011). On the one hand, the participation of men in female-dominated occupations (such as counselling, nursing and elementary teaching) is increasing (Perra & Ruspini, 2013). On the other hand, we are witnessing an increase in men's household work and/or care work contribution and a growing fathers' involvement in their children's lives (see, e.g. Dermott, 2008; Doucet, 2006; Oláh *et al.*, 2018; Parker & Livingston, 2018). Being a good parent seems to be one priority in the lives of millennials and young men today express a strong need and desire to be involved parents (Parker & Livingston, 2018). The changing role of fathers brings fresh challenges, as these changes are not without tensions (Ruspini, 2019). A study conducted in 2016 by the Boston College Center for Work and Family (Harrington *et al.*, 2016) – aimed at better understanding the ways in which millennial fathers view parenting, careers and work–family balance[3] – well summarises those tensions. For the majority of millennial fathers there has been significant movement towards greater gender equality and the need to find a way to share more equally in caregiving. At the same time, the study shows that traditional gender roles and values continue to exist. Millennial fathers perceive that their workplace cultures encourage thinking that includes the ideas that work should be 'priority number one', that the ideal employee is available 24/7 and that good employees work long hours. Most men face complex conflicts in trying to find and maintain a balance between work and parenting, between the desire to be engaged fathers and to pursue a career (Crespi & Ruspini, 2016). Millennial fathers are thus likely to experience similar levels of work–family conflict to millennial mothers. Not surprisingly, workers' perspectives and expectations have also changed: previous studies paint a fairly consistent picture of a general increase in the importance of and need for work–life balance with successive generations (Buzza, 2017; Insead, Head Foundation and Universum, 2014; Lyons & Kuron, 2014). As the roles of fathers grow, so do expectations and information needs. A large number of millennial fathers seek parenting information online, turning to the internet – not just family and friends – for fatherhood tips and guides (Pecorelli, 2016). As anticipated in Chapter 2, the fact that millennial parents consider having children central to their identity has an impact on tourism choices. Millennial parents do not hesitate to travel with their children and take more trips, and longer trips, compared to older generations. For example, the 'MMGY Global Portrait of American Travellers'[4] explains that millennials are driving multigenerational travel: millennial families travel much more than couples or singles. Millennials with children are also travelling more internationally than any other generation. The survey conducted by the Boston Consulting Group on US millennials[5] (Barton *et al.*, 2013) reveals that millennials often travel with spouses and children and they are also more likely than non-millennials to travel for leisure in organised groups, with extended families, or with adult friends. The online 'Millennial and Gen Z Traveller Survey 2019' by Skift Research (Carty, 2019) – that collected responses from travellers aged 16–38 who reside in five major travel markets: Australia, China, India, the US and the UK[6] – shows that millennial parents are more interested in many types of trips and trip activities, and more concerned about environmentally responsible travel. Finally, the 'MMGY 2019–2020 Portrait of American Travellers' survey paints a picture of rapidly changing priorities among travellers, driven by growing concerns over safety, the quick adoption of the new sharing economy and a more conscientious approach to travel.

A clear trend is the increased confidence in technology and the trust in robots and artificial intelligence (AI) among the younger generations (see Chapter 3 for details). The 2017 Deloitte Millennial Survey[7] shows that many recognise the benefits of automation in terms of productivity and economic growth; they also see that it provides opportunities for value-added or creative activities, or learning new skills (Deloitte, 2017). The 'Generation AI 2019: Third Annual Study of Millennial Parents of Generation Alpha Kids Second Annual Generation AI Survey' conducted by IEEE (2019) – that surveyed 2,000 parents, aged 23–38 years old, with at least one child 9 years old or younger: 400 each in Brazil, China, India, the US and the UK – reveals the confidence millennial parents may have in using AI and emerging technologies for the health and wellness of their Generation Alpha children. Because millennials are more acquainted with emerging technologies, including robotics, it would become possible for the tourist accommodation sector to introduce those technologies in the near future to provide entertainment and enhance the guest experience.

Past, Present, Visions of the Future

The aim of this chapter was to discuss, using a generational approach, the nexus between gender, generational change and tourism.

Traditional gender ideologies that emphasise the value of and the need for distinctive roles for women and men have severely limited women's mobility and travel experiences through history. In the 19th century, thanks to significant improvements in women's education, the feminist movements, women's activism and a number of changing social conditions, economic factors and technological innovations, women increasingly took part in travel. Women's travel and tourism has increased exponentially among younger generations. As said, female solo travel is on the rise globally: women are increasingly choosing to travel independently, without the companionship of partners (male or female) or the support of packaged trips or tours. For example, the 2018 British Airways 'Global Solo Travel Study' – that surveyed almost 9,000 18–65 year olds across Brazil, China, France, Germany, India, Italy, the UK and the US – found that over 50% of women have taken a holiday by themselves, with 75% of women planning a solo trip in the next few years. In Italy, 63% of women interviewed have explored another country alone, followed by Germany at 60% (British Airways, 2018). This trend has been coupled with the rise of global travel communities, such as 'The Solo Female Traveller Network', an online community connecting women of all ages; 'We Are Travel Girls', aimed at inspiring, connecting and empowering women travellers around the world; 'She Roams Solo', a travel social network whose aim is to help inspire, educate and support women on their travelling endeavours.

It is within this context marked by progress towards gender equality, where more and more women are challenging gender stereotypes, increasing their autonomy and agency and making specific travel choices (UNWTO, 2019), that the Covid-19 pandemic emerged. The outbreak of the coronavirus disease in 2019 dramatically changed our lives, increasing, at the micro level, isolation, anxiety and fear, and at the macro level, economic insecurity and job instability. Covid-19 is undermining trust in government and institutions and people are much more conscious of how they interact with other people. Due to lockdowns, mobility and travel restrictions, border closures and the confinement of people, cities and states, tourism has been and continues to be one of the sectors hardest hit by the Covid-19 crisis (UN Women, 2020a; UNWTO, 2020a, 2020b, 2020c). All 2020 forecasts for international arrivals and receipts have been radically revised downward and all predictions about the future of tourism need to be re-evaluated. According to UNWTO (2020b), international tourist arrivals (overnight visitors) declined 65% in the first half of 2020 over the same period in 2019 year, with arrivals in June down 93% (UNWTO, 2020c). The massive drop in international travel demand over the period January–June 2020 translates into a loss of 440 million international arrivals and about $460 billion in export revenues from international tourism. The pandemic crisis is putting millions of jobs in the global travel and tourism sector at risk, with young people and women particularly vulnerable to job loss.

Millennials are likely to be among the ones who suffer most, not only because they are poorer than members of previous generations but also because many of them have formed their own families and have children. As written (The Council of Economic Advisers, 2014), one of the most important markers for millennials is that they have come of age during a very difficult time in the global economy (Martin & Lewchuk, 2018). They are less well off than members of earlier generations when they were young, with lower earnings, fewer assets and less wealth. Gen Zers are roughly 23 and younger and the majority have not yet started their careers, so it is not easy to predict how the pandemic will affect this generation worldwide. Unlike millennials, this young generation was in line to inherit a stronger economy with lower unemployment. That has all changed now, as Covid-19 has reshaped the social, political and economic global landscape picturing a very uncertain future (Parker & Igielnik, 2020). For Gen Zers, the effects of the economic downturn will likely be felt for years, as they are either still in school or just trying to start their careers. Teens are facing a very uncertain future and this can negatively affect relationships. A recent survey by Fuse (2020) on 1000 or 1,000? of the older members of Gen Z has revealed that their anxiety and depression due to social distancing and self-isolation have increased, as well as their consumption of social media.

However, in the face of unprecedented health and economic adversity caused by the Covid-19 pandemic, millennials and Gen Zers also express a vision to build a better future. The '2020 and 2021 Deloitte Global Millennial Surveys' (Deloitte, 2020, 2021)[8] reveal that both generations remain resilient, share a vision for a better future and are determined to help drive positive change both locally and globally. The 2020 study also explains that despite the individual challenges and personal sources of anxiety, millennials and Gen Zers have remained focused on larger societal issues, both before and after the onset of the pandemic. The Covid-19 crisis may have reinforced these inclinations, as nearly three-fourths said the pandemic has made them more sympathetic to the needs of others, and that they will take action to positively impact their communities. Climate change emerged as a critical issue for millennials and Gen Zers both before and during the Covid-19 crisis. A vast majority (80%) also think governments and businesses need to make greater efforts to protect the environment, yet they are concerned that the economic impact of the pandemic might make this less of a priority.

As for gender, there is a growing concern that the ongoing pandemic crisis will disproportionately affect women and girls because of existing gender inequalities that can be exacerbated and increase the risk of discrimination and violence against women. Women from vulnerable groups (such as women with heavy care responsibilities, single mothers, migrant women, domestic workers, women with precarious work, women from rural communities) will be faced with a much higher economic cost than men (Durant & Coke-Hamilton, 2020; UN Women, 2020a, 2020b; UNFPA, 2020). The application of confinement measures that seek to protect public health and prevent the collapse of health services is, in fact, not gender neutral (OAS–CIM, 2020). The first consideration to make is that the pandemic is heavily increasing women's burden of unpaid care work. Since the health emergency caused by Covid-19, women have disproportionately taken up the role of primary carer without (or with much reduced) institutional support. Mass closure of schools, childcare centres and centres for elderly and disabled people has particularly affected women because they still bear much of the responsibility for providing care. Women are particularly challenged by the need to simultaneously manage paid work (either at home or outside, both online and offline) and their usual, unpaid domestic workload. The situation is more difficult in families where parents are trying to work from home while caring for both their children and other relatives. Millennial women seem to be specifically challenged because they are now at the age where they have young children: many women who were struggling with childcare responsibilities pre-Covid-19, are now going to face additional barriers. While the pandemic will eventually come to an end, the fear is that it could have a long-lasting impact on young women's careers. The changing nature of work, with the expansion of atypical forms of employment, indeed poses further challenges: flexible working time brings about both work intensification and work extensification (Lu, 2009).

In regard to employment, women are more likely than men to lose their jobs because they remain clustered in secondary labour markets marked by uncertainty, precariousness and vulnerable job positions (Durant & Coke-Hamilton, 2020; UN Women, 2020; UN, 2020a, 2020b). Women are also disproportionately represented in industries that are expected to decline the most due Covid-19. The McKinsey Global Institute (Magvadkar *et al.*, 2020) estimates that 4.5% of women's employment is at risk in the pandemic globally compared with 3.8% of men's employment: the reason is that women have more than the average share of employment in the most affected sectors such as accommodation and food service. As just said, the tourism sector is affected more than others by the current pandemic, and vulnerable groups are among the hardest hit (UNWTO, 2020d). As a sector with a majority female workforce worldwide (54%: UNWTO, 2019) and most women in low-skilled or informal work, women will feel the economic shock to tourism caused by Covid-19 quickest and hardest. The lack of legal protections inherent to informal employment (such as crafts, food and beach vendors) leaves women particularly exposed to the crisis. Likewise, women with their own small businesses are at risk.

Another key area of concern is women's health, well-being and GBV. Women are particularly exposed to the risks to health and life posed by the pandemic, with many on the frontlines in the Covid-19 fight, providing essential medical and other services. Women are disproportionally represented in the health and social services sectors. They constitute 70% of the workers in these sectors globally, and this substantially increases their risk of exposure to the disease, especially if they do not have adequate protection against transmission (UN Women, 2020b). In some countries, Covid-19 infections among female health workers are twice that of their male counterparts (UN Women, 2020a). When mobility is restricted, people are confined and protection systems weakened, women and girls are at greater risk of experiencing GBV and abuse (OECD, 2020; UNFPA, 2020). Sexual harassment and other forms of GBV are widespread in the tourism industry: women and girls encounter multiple and interconnected manifestations of violence in tourism consumption and production, such as physical, sexual, emotional or socioeconomic abuse (Vizcaino-Suárez *et al.*, 2020). There is thus a strong need to strengthen legal protections against domestic violence and other forms of GBV, that will make tourism a safer space for women workers as the sector recovers.

Women can help increase resiliency, support a more rapid recovery and achieve a more inclusive, responsible and sustainable economy. Women are indeed playing an outsized role responding to Covid-19, both in the private and public spheres: as mothers, daughters, nurses, midwives and community health workers they are contributing to building the resilience of the communities most deeply affected by Covid-19. There is a need for forward-looking socioeconomic plans and policies aimed at empowering girls and women, also by including both women and women's organisations at the heart of the Covid-19 global response (UN Women, 2020b). The 'end of global travel and tourism as we know it' may represent a golden opportunity for making the right moves towards more responsible and balanced tourism which cannot fail to address the issue of gender equality and women's empowerment, vital elements to reach sustainability. The commitment that tourism must make in this sense is multidirectional: promoting women's participation in tourism education and training, reducing barriers to entry, supporting women's inclusion in tourism planning and management, allowing work–life balance, improving the practice and perception of personal safety and increasing protection against GBV (UNWTO, 2020d).

Notes

(1) Socialisation is the process of transferring norms, values, beliefs and behaviours to future group members and gender socialisation is the process of educating and instructing potential men and women how to behave as members of that particular group.

(2) Survey of 10,682 US adults conducted using Pew Research Center's American Trends Panel (ATP) between 24 September and 7 October 2018, and a survey of 920 teenagers between 13 and 17 conducted using the NORC AmeriSpeak panel between 17 September and 25 November 2018, for a combined sample size of 11,602.
(3) This study surveyed individuals aged 22–35 with at least two years' professional work experience who were employed at one of five large global corporations. The study included both a quantitative and a qualitative element. The quantitative survey was conducted online. While all of the businesses were global in scope, the survey was only administered within their US operations. A total of 1,100 employees completed the survey across five companies. Employees' participation in the study was voluntary. For this exploration of millennial fathers, the study draws mainly from the responses given by 33% of the study subjects (327 participants) who were parents and especially the 151 fathers.
(4) The 'MMGY Global Portrait of American Travellers' study is based on a yearly survey of nearly 3,000 American adults who have taken at least one trip of 75 miles or more over the past 12 months.
(5) The survey was conducted on 4,000 millennials (ages 16–34) and 1,000 non-millennials (ages 35–74) in the US. The survey included questions about preferences related to business and leisure travel.
(6) The study 'Millennial and Gen Z Traveller Survey 2019' surveyed respondents ages 16–38 who reside in five countries around the world. Around 1,000 responses each were collected from China (1,143), India (1015) and the US (1,046), due to the larger size of these markets, and around 500 each were collected from Australia (523) and the UK (509). To qualify, respondents had to indicate that they had taken at least one leisure trip in the last 12 months.
(7) The '2017 Deloitte Millennial Survey' questioned almost 8,000 millennials across 30 countries. Participants were born after 1982 and represent a specific group of this generation: those who have a college or university degree, are employed full-time and work predominantly in large, private-sector organisations.
(8) The 2020 report is based on two sets of surveys. The first began prior to the Covid-19 outbreak using an online interview. Fieldwork was completed between November 2019 and January 2020 and 13,715 millennials across 43 countries and 4,711 Generation Z respondents from 20 countries were surveyed. A second survey was conducted in a similar fashion between April and May 2020, during the worldwide pandemic: it questioned 5,501 millennials and 3,601 Gen Zers over 13 countries. Millennials included in the study were born between January 1983 and December 1994. Gen Z respondents were born between January 1995 and December 2002. The overall sample size of 27,500 represents the largest survey of millennials and Gen Zers completed in the nine years Deloitte Global has published this report. No respondents in the former survey were queried in the latter: https://www2.deloitte.com/global/en/pages/about-deloitte/articles/millennialsurvey.html. The 2021 report collects the views of 14,655 millennials and 8,273 Gen Z members (22,928 respondents in total) from 45 countries across North America, Latin America, Western Europe, Eastern Europe, the Middle East, Africa and Asia Pacific. The survey was conducted using an online, self-complete style interview. Fieldwork was completed between 8 January and 18 February 2021: https://www2.deloitte.com/content/dam/Deloitte/global/Documents/2021-deloitte-global-millennial-survey-report.pdf.

6 LGBTQ+ and Next-Gen Tourism

Fabio Corbisiero

Introduction

This chapter examines the relationship between lesbian, gay, bisexual, transgender, queer and others (LGBTQ+) people and tourism, specifically focusing on the younger generations (millennials and members of Generation Z). This chapter aims to integrate a sociological perspective into the generational analysis of tourism processes. The sociological perspective essentially conceptualises LGBTQ+ issues as a 'community' referring to a broad coalition of groups that are diverse with respect to gender, sexual orientation, race/ethnicity and socioeconomic status.

Thus, while this chapter focuses on the community that is encapsulated by the acronym LGBTQ+, it is important to highlight the importance of recognising that the various populations represented by 'L', 'G', 'B' and 'T' are distinct groups, each with its own special social-related concerns and needs. Not forgetting that individual and societal understandings of sexuality and gender identity are constantly changing. The concept of LGBTQ+ expresses the varied system of beliefs and sexualities whereby different cultures and identities determine the functions and responsibilities of each sexual identity within the acronym LGBTQ+ (Plummer, 1981).

Even though the study of LGBTQ+ generations in leisure mobility is fairly recent, it represents a key issue in the late-modern development of tourism research, strongly linked to the generational turnover (Corbisiero & Ruspini, 2018).

Indeed, LGBT tourism contributes to enhancing the visibility, role and recognition of homosexual people, and benefits destinations by associating their brand image with inclusiveness and diversity. The continuous expansion experienced by this segment nowadays reflects the great array of opportunities it presents for tourism stakeholders and destinations. In order to correctly harness the existing potential within this area, there is a need to overcome the challenge of correctly understanding and adapting the tourism offer to the preferences and needs of the diverse LGBTQ+ travellers (Corbisiero & Monaco, 2022, in press).

Despite this growing attention and major changes in the laws and norms surrounding the issue of same-sex marriage and the rights of homosexual people around the world, LGBTQ+ issues seem to be underrepresented in tourism studies including several social sciences fields. Lately, certain 'niche' scientific segments such as gender and sexual identities in tourism have received sufficient research attention from the social sciences, but due to its complexity homosexual tourism remains a neglected area of scholarly exploration (Monterrubio, 2018).

From this point of view, there is also a geographical perspective that is linked to how high the degree of inclusiveness of homosexuals is in the geographic areas where tourism scholars operate. It may take some areas more time to begin welcoming LGBTQ+ people.

The 2019 survey by Pew Research Center (PRC) shows that while majorities in 16 of the 34 countries surveyed say homosexuality should be accepted by society, global divides remain. Whereas 94% of those surveyed in Sweden say homosexuality should be accepted, only 7% of people in Nigeria say the same. Across the 34 countries surveyed, a median of 52% agree that homosexuality should be accepted with 38% saying that it should be discouraged. On a regional basis, acceptance of homosexuality is highest in Western Europe and North America. Central and Eastern Europeans, however, are more divided on the subject, with a median of 46% who say homosexuality should be accepted and 44% saying it should not be accepted. In sub-Saharan Africa, the Middle East, Russia and Ukraine, few say that society should accept homosexuality; only in South Africa (54%) and Israel (47%) do more than a quarter hold this view. Again according to the 2019 PRC survey, more than three quarters of those surveyed in Australia (81%) say homosexuality should be accepted, as do 73% of Filipinos. Meanwhile, only 9% in Indonesia agree. In the three Latin American countries surveyed, strong majorities accept homosexuality in society.

The generational variable seems to have a positive influence on tolerance towards homosexual people. In some statistics it appears that younger generations in several countries view homosexual subjects more tolerantly. For example, an online survey of 37,653 respondents in 39 countries carried out by the PRC in 2013 found that younger respondents were consistently more likely than older respondents to say homosexuality should be accepted by society. The survey finds that the attitudes and experiences of younger adults in the LGBTQ+ population differ in a variety of ways from those of older adults, perhaps a reflection of the more accepting social milieu in which younger adults have come of age. For example, younger gay men and lesbians are more likely to have disclosed their sexual orientation somewhat earlier in life than have their older counterparts. Age differences are particularly evident in the Republic of Korea, Japan and Brazil, where those younger than 30 were found to be more accepting than those aged 30–49 who, in turn, are more accepting than those aged 50 and older. Furthermore, the research found that Mexicans and Chinese aged 18–29 are more likely to be accepting than those in each of the older groups, while in the Russian Federation, El Salvador and Venezuela, those younger than 30 are more tolerant of homosexuality than are those aged 50 and older. Among the more mature outbound markets of North America, acceptance levels among millennial respondents were much higher: 90% in Spain, 87% in Germany, 87% in Canada, 86% in Italy, 79% in the UK and 70% in the US, each consistently showing higher levels of acceptance than older age groups. Attitudes on the acceptance of homosexuality are shaped by the country in which people live. Those in Western Europe and the Americas are generally more accepting of homosexuality than those in Eastern Europe, Russia, Ukraine, the Middle East and sub-Saharan Africa. And citizens in the Asia-Pacific region are generally split.

This generational shift in attitudes is likely to be a function not only of the economic development of nations, but also of religious and political ideologies. This shift seems to have important consequences for the future development of LGBTQ+ travel processes because those world areas that are seeking to reach this young segment will have to find innovative ways of making their destinations attractive to an audience that feels less defined by their sexuality, or the need to hide it from wider society, while at the same time recognising the need to welcome those who feel different from the mainstream and want to express this through their travel choices. With millennials continuing to dominate the workforce and Generation Z beginning their college journey, it is crucial to understand this new wave of LGBTQ+ travellers. Tourism studies are now called upon to focus on this burgeoning segment because of changing social attitudes and the realisation that the key to success is not mere mass-marketing but rather a strong 'niche position' (Novelli, 2005) that offers something singular to a particular set of consumers. One thing is for certain, and that is that both millennials and Gen Zers are comfortable with disrupting the norm. They are highly connected, technologically advanced and globally conscious and far more open to trying out new products and concepts than their parents or grandparents ever were. This chapter focuses on millennial and Gen Z LGBTQ+ travellers through a review of theoretical studies and empirical research and discusses 'rainbow touristification' (Corbisiero, 2016).

We will be able to argue that as a social phenomenon, millennials and Gen Z LGBTQ+ tourism is not only a matter of being young but it is also a tool against social, political and economic inequalities (Murray, 2007: 49). Social research has proven (Hartal, 2020; Monaco, 2018; Ram et al., 2019) that when progress is made towards equal rights (e.g. the introduction of same-sex marriage), destinations benefit from a boost to their brand and increased arrivals and spending associated with the social needs of the youngest generations.

From Grand Tour to the Present: The Rainbow Travel

Since the first and most potent protest staged to secure full citizenship, the 1969 Stonewall Inn raid on Christopher Street in New York, LGBTQ+ people have progressively protested against a heteronormative world, making claims for non-discrimination and inclusive spaces through their mobility and travel as well. From that moment on, North America and later many democratic countries all over the contemporary world have committed to assuring homosexual citizens equality and increasing attention to the specific issues of gay travellers. An important milestone for the gay community was the depathologisation of their identity in the early 1970s when the American Psychiatric Association removed homosexuality from their list of mental disorders. Much later, in 1990, the World Health Organisation also eliminated homosexuality as a mental illness from the 'International Classification of Diseases' (Drescher, 2015).

Since then there has been a progressive trend towards ever-wider legal and political opportunities for gender and sexual minorities with the rise of positive social attitudes regarding equal rights for homosexual people, resulting in socioeconomic benefits in a range of sectors including tourism (Guaracino & Salvato, 2017). This trend has involved the birth of more effectively managed 'gay-sensitive' tourist activities. For instance, the 1983 founding of the International Gay and Lesbian Travel Association (IGLTA) served to support the development of homosexual tourism and highlight its social and economic importance. Today, the IGLTA is the most important organisation in the homosexual tourism sector and it has numerous affiliates among various operators in over 70 countries around the world.

Travel by homosexual people is not a recent phenomenon, as they have been setting out for centuries in search of 'safe Arcadias' (Aldrich, 2004) where homosexuality was not ostracised as much as it was in their home countries. For instance, in his examination of sexually motivated travel to the Mediterranean, Robert Aldrich (1993) has found that many gay men in the late 19th and early 20th centuries shared a 'cultural and political creed' that Aldrich defines as the 'Mediterranean myth'.

The connection between travel and sexuality was first identified by social scientists with regard to European colonialist erotic-exotic journeys. Littlewood (2001) describes such travellers as 'sexual pilgrims' and identifies homosexual men as a significant contingent among them. The 'Homosexual Grand Tour' has been studied as part of the bourgeois practice of touring, as well as the 'literary expatriation' from a homophobic motherland such as England, France or Austria to a tolerant

Mediterranean area such as Italy. This was a period in which tourism was not yet as formalised and institutionalised as it is today. While its main purposes were cultural and sociopedagogical, the Grand Tour also represented a recreational opportunity and an affirmation of the traveller's social status (De Seta, 1993). During that period of travelling to discover Europe, (Grand) tourists came into contact with the Mediterranean culture and the vast historical-artistic heritage of classicism found in the Mediterranean area (Hersant, 1988). This experience also represented a means for aristocrats to send their children or relatives whom they considered 'inverts' because of their homosexual orientation away from home. In other words, with the excuse of study and education, the relatives of young homosexual men were able to send them out of their home country in the hope that the trip would somehow make them 'come to their senses' (Fagiani, 2010).

At the same time, however, many young gay men of the time also experienced the Grand Tour with the hidden intention of being able to live their identities more freely away from home (Clift & Wilkins, 1995). These travellers stayed mainly in Italy and Greece, as these more than other Mediterranean countries were able to offer young homosexual men spaces of freedom and self-realisation. Unlike many Northern European countries of the period, most Italian states lacked anti-homosexual legislation or the laws were not enforced (Beccalossi, 2015). Several reasons made Italy, especially the South, so appealing to British or North-European homosexuals, some of which were eminently practical. In southern Italy throughout the 19th century, male same-sex acts were not legally punished and, from 1889, through the enactment in law of the 'Zanardelli Code', male same-sex acts were decriminalised on a national scale. Prosecution at home encouraged wealthy British men to quit their own country and go into exile or, if they could not leave permanently, to make frequent visits to Italy in order to take advantage of Italian tolerance towards same-sex desires. Moreover, in Italy it was possible to live well on less money than in the rest of Europe.

According to the IGLTA, homosexual tourism began to be recognisable as a specific tourist segment thanks to the development and marketing of tourism products and services specifically designed for homosexuals, such as facilities, accommodation and amenities. Therefore, the creation of publications targeting gay male travellers was probably the earliest recognition of this population as a market segment with its own specific interests and needs. The first of such examples was the 1960s gay men's travel guide *The Damron Address Book* that published an annual directory listing establishments frequented by and friendly to gay men (Sosa, 2019). Another important supplementary factor in granting increasing visibility to homosexual tourism has been the community's participation in global mobility linked above all to the civil rights movements and developments in gay liberation that took place from the 1960s onwards in major North American and European cities whose liberal climates attracted people of all sexual orientations and gender identities. Rainbow cities such as New York, Chicago and San Francisco became synonymous with the homosexual rights movement, but they also triggered the mobility of large groups of gay travellers, causing a butterfly effect even in large European cities such as London and Paris. Later, beach-side communities such as Provincetown or Key West in the US and Ibiza, the Gran Canarias and Mykonos in Europe supported the growth of homosexual tourism, so much so that homosexual travellers are currently one of the key target markets of the world tourism industry.

Why do we need to talk about LGBTQ+ tourism? While data are continually sought to explain these trends, LGBTQ+ travellers have become recognised as a segment that travels with greater frequency and demonstrates higher than average patterns of spending. Social change and improvements in the legal recognition and protection of LGBTQ+ people around the world have meant that LGBTQ+ tourists have gradually become more visible and with this, more easily targeted as a 'consumer segment'. Hughes (1997) argued that vacations, especially holidays spent abroad, offer homosexuals, especially gay men, the opportunity to experience LGBTQ+ cultures, practices and lifestyles that might not be readily available in their homelands. Above all in the Western perspective on homosexuality, travelling abroad has frequently been understood as a personal journey of self-discovery and self-expression in new spaces: learning about the ways of others is ultimately a means to discover and understand oneself (Hughes et al., 2010).

Over time, travel has taken on different forms and meanings for homosexual people (Vorobjovas-Pinta, 2021). The ways in which this specific form of mobility takes place is the result of a series of factors that go beyond simple personal inclinations. In fact, while heterosexual people in contemporary society are ideally free to travel following their inclinations and passions, matters are still somewhat different for the homosexual niche (Moreira & Campos, 2019; Stuber, 2002) in that their reception and treatment varies from one local area to another. Too often, homosexual people continue to face prejudice and discrimination. Depending on the character of the tourist destination, therefore, homosexual travellers might be likely to conceal their sexual orientation when travelling or, in gay-friendly destinations, feel encouraged to come out (Friskopp & Silverstein, 1996).

Fortunately, in recent years, many tourist destinations have taken effective measures to cope with discrimination. These include removing criminal sanctions for consensual same-sex conduct; legal prohibition of discrimination on the basis of sexual orientation, gender identity and intersex status; legal recognition of the gender identity of transgender persons without abusive requirements; legal recognition of same-sex couples and their families; protections for the physical integrity of intersex children; public education and awareness-raising campaigns to combat homophobia and transphobia; establishing shelters for homeless LGBTQ+ and intersex youth; and anti-bullying initiatives in schools. There is still a long way to go before we can claim that the full inclusion of homosexual citizens in general and their right to mobility in particular are guaranteed throughout the world.

What Kind of Travellers are LGBTQ+ People?

Surely everyone wants to feel welcome, safe and respected while travelling. This kind of 'welcome feeling' has been growing extremely rapidly and favourite destinations for the homosexual community have emerged in that way, especially in Western European countries and North America. Recently coined terms such as 'gaycation' (Collins English Dictionary, 2014) referring to a version of a vacation that includes a pronounced aspect of homosexual culture and branding in either the journey or the destination and 'dual income, no kids' (DINKS) (van Gils & Kraaykamp, 2008) are sociological indicators of this growing segment. In particular, researchers have focused on studying tourist destinations and other locally rooted tourist consumption–production complexes (Costa & Lopes, 2013; Zukin, 1995), but these phenomena have also come under critique: for instance, the alleged economic benefits of 'gay-friendliness' have given rise to official pink tourism marketing strategies and new segments of (homo)capitalism with mostly white gay men being 'colonised' by the market (Nast, 2002). Tourism bolsters the homosexual lobby and, embedded in cultural proximity, such travel may even be on its way to displacing heteronormative capitalist patriarchies (e.g. Knopp, 1992).

An overview of the existing empirical research on homosexual people as tourists reveals that a considerable body of literature was published in the 1990s and early 2000s concentrating on issues such as the economic power of (especially gay men) homosexual travellers and their supposed 'pink dollar' spending capacity (e.g. Holcomb & Luongo, 1996) as well as homosexual people's preferred destinations and motivations for travelling (e.g. Kinnaird *et al.*, 1994; Pritchard *et al.*, 2000; Puar, 2002). The predominance of the social research based on the male gender has often grouped LGBTQ+ persons together, in false conceptualisations of homogeneous tourism processes. A misconception that has been debunked by the diversity within the community (Therkelsen *et al.*, 2013; Corbisiero & Monaco, 2017). For decades, the touristic theme, within the sexual identity topic, was dominated by concepts like 'sexual' and 'male'. As if to say that the LGBTQ+ tourism literature has been relatively homogeneous over time in terms of the topics covered, but not in terms of individual sexual identities. Noticeably, non-binary and transgender identities continue to be underrepresented. Transgender individuals' travel motivations and tourism experiences have been virtually ignored. Yet, many of the assumptions made by scholars are equally extendable to the entire homosexual community.

As Hughes's (2002) study pointed out, homosexual travellers carefully evaluate the normative and cultural conditions of a potential destination in relation to homosexuality before organising a trip. His research reveals that homosexual people grant little or no consideration to the idea of visiting geographic areas such as Jamaica or China or African, Muslim or Arab countries. These data are in line with a more recent study finding (CMI, 2014) that only a minimal percentage of members of the homosexual community (11%) look to countries with anti-homosexual laws (such as Russia, Jamaica or Kenya) as potentially interesting tourist destinations. Other common topics of social scientific research on homosexual tourism are sexuality and holiday choices (e.g. Blichfeldt *et al.*, 2011; Casey, 2009), such as the configuration of homosexual leisure spaces (Pritchard *et al.*, 2002). In addition, some scholars have analysed the holiday profiles of older gay men (e.g. Hughes & Deutsch, 2010) and the destinations' competitiveness in the gay tourism segment (Melián-González *et al.*, 2011) thanks above all to the proliferation of LGBTQ+ tourism guides. As we have argued, social researchers have failed to differentiate between gay and/or lesbian tourism experiences and those of non-binary or trans individuals, mistakenly reporting all these specific groups' holiday experiences as defined by sexual issues. Fiani and Han (2019) recently asserted that the 'T' in 'LGBTQ' (i.e. LGBT and queer or questioning) has been rendered silent within tourism studies. Some recent studies about trans travellers have emphasised the importance of affirming the non-conforming gender dimension during their own travel, rather than that of sexual identity as for homosexuals (Monterrubio *et al.*, 2020; Olson & Reddy-Best, 2019). In relation to lesbians, neglecting them as a sociological category in tourism studies may have to do with their perceived reduced market power and their socialisation patterns, at least in some countries. According to Hughes (2006), lesbians have not been considered economically powerful or visible and have not been targeted as a separate consumer group. They are also considered to be more difficult to reach as they are less concentrated in urban areas, less likely to socialise in LGBTQ+ clubs or events and are more oriented towards private social activity and entertainment. It was Hughes (2006) himself who suggested that lesbian women are less interested in visiting LGBTQ+ destinations because they would not feel comfortable within the patriarchal domain of homosexual males, typical of some gay destinations. Going beyond this critical dimension of the lack of distinction of LGBTQ+ people's behaviours, other common topics are sexuality and holiday choices (Blichfeldt *et al.*, 2011; Casey, 2009; Hughes, 2002), gay and lesbian tourist experiences (Poria, 2006) and the configuration of homosexual leisure spaces (Pritchard *et al.*, 2002).

In addition, many scholars have analysed the holiday profiles of older gay men (Hughes & Deutsch, 2010) and the destinations' competitiveness in the gay tourism segment (Melián-González *et al.*, 2011).

LGBTQ+ tourists have also been categorised as 'trendsetters', who search for a gay social life, culture, sights, comfort and relaxation, have the desire to distance themselves from routines and aim to express their sexuality in safe environments, especially 'closeted' LGBTQ+ people who search for the opportunity to get free from societal constraints. Further motivations to travel are the needs for escape, a sense of belonging and safety and the opportunity to live one's identity more freely than in the larger heterosexual world, in which their sexuality is ridiculed or hidden (Pritchard *et al.*, 2000). Another American study conducted on a sample of over 5,000 LGBTQ+ respondents residing in all 50 states of the US reports among the main travel motivations: 'vacation/ leisure', 'work', 'culture', but also 'family holidays' and 'honeymoon', with an average of 10 nights spent in accommodation facilities (CMI, 2019).

Although the demographic, motivational, behavioural and identity aspects of 'pink travellers' may be more complex than this data framework describes, a great deal of the literature on LGBTQ+ tourism is now outdated and therefore requires a renewed and empirical assessment of the phenomenon and of its ongoing changes. Perhaps more indicative of the appeal of the research on homosexual lives is the increasing recognition of (homo)sexuality as a salient category of difference within travellers' behaviours. In paying attention to gay and lesbian lives, lifestyles, markets, holiday destinations and scenes, researchers often assume particular commonalities, or at least communities (Corbisiero, 2016; Valentine & Skelton, 2003). Following from this, there is a related assumption (that can implicate all sexualities and often goes unquestioned) of a pre-given identity or behaviour that can be objectively measured and sociologically studied.

The Generation Lens to Observe LGBTQ+ Tourism

As we recalled in Chapter 1, generational analysis in the tourism literature has decisively increased. Studies have identified that the values and beliefs that are dominant among members of different generations can have a significant impact on tourism processes and trends. The proliferation of the generational literature within this particular field has been rather effervescent in the last few years. Most of the recent tourist studies focusing on generations are comparisons between millennials, Generation Y and Z, Net Generation and Gen Alpha; less frequently they focus on generational groups born before the global spread of the internet, now firmly entered in the international literature, or single people (Li *et al.*, 2013; Xiang *et al.*, 2015). Recently, another generation – dubbed the

Table 6.1 Types of modern generations

Generation name	Births – start	Births – end	Youngest age today*	Oldest age today*
Baby boomer generation	1946	1964	57	75
Generation X (Baby Bust)	1965	1979	42	56
Xennials	1975	1985	36	46
Millennials, Generation Y, Gen Next	1980	1994	27	41
iGen/Gen Z	1995	2012	9	26
Gen alpha	2013	2025	1	8
Coronnials	2020	–	1	2

*Age as of 2021.

'coronnials' (Harmony, 2020) – has also been added to this generational list, indicating not only all those born during the Covid-19 pandemic but also, in more general terms, human resilience in a time of great unknowns (Table 6.1).

The use of generational lenses to analyse social behaviour is a useful tool in the sociological research about LGBTQ+ people (Russell & Bohan, 2005). The radical discrepancy between the life of today's LGBTQ+ youths and that of their elders when they were young, the pervasive age-emphasis within LGBTQ+ communities, the extreme speed of change that renders today's certainties tomorrow's irrelevancies, can contribute to providing a clearer description of LGBTQ+ tourism.

Generational differences allow to intercept social changes in LGBTQ+ communities and to highlight the break with the past and to hypothesise possible future scenarios of this population. Research by Williams Institute showed that millennials are significantly more likely to openly identify as LGBTQ+ than older generations. This for sure can be attributed to increasingly accepting social environments, cultural and media changes and overall a widespread sense of social safety especially in urban contexts. Additionally, increased LGBTQ+-friendly tourist destinations has led to a greater grasp of LGBTQ+ communities which has diminished stereotypes and stigma, making it generally less challenging for many young people to travel around the world openly identifying as LGBTQ+ or queer (Figure 6.1).

Given that millennials are more likely to openly support LGBTQ+ than generations before them, millennials are also much more likely to be allies of LGBTQ+ communities. Still, some scholars argue that millennials and Gen Z are more apt to describe fluid or multiple identities than older generations (Gardiner et al., 2013). Some of these beliefs are also explained by the authors of the special issue 'Millennials and Generation Z: Challenges and Future Perspectives for International Tourism' (Corbisiero & Ruspini, 2018).

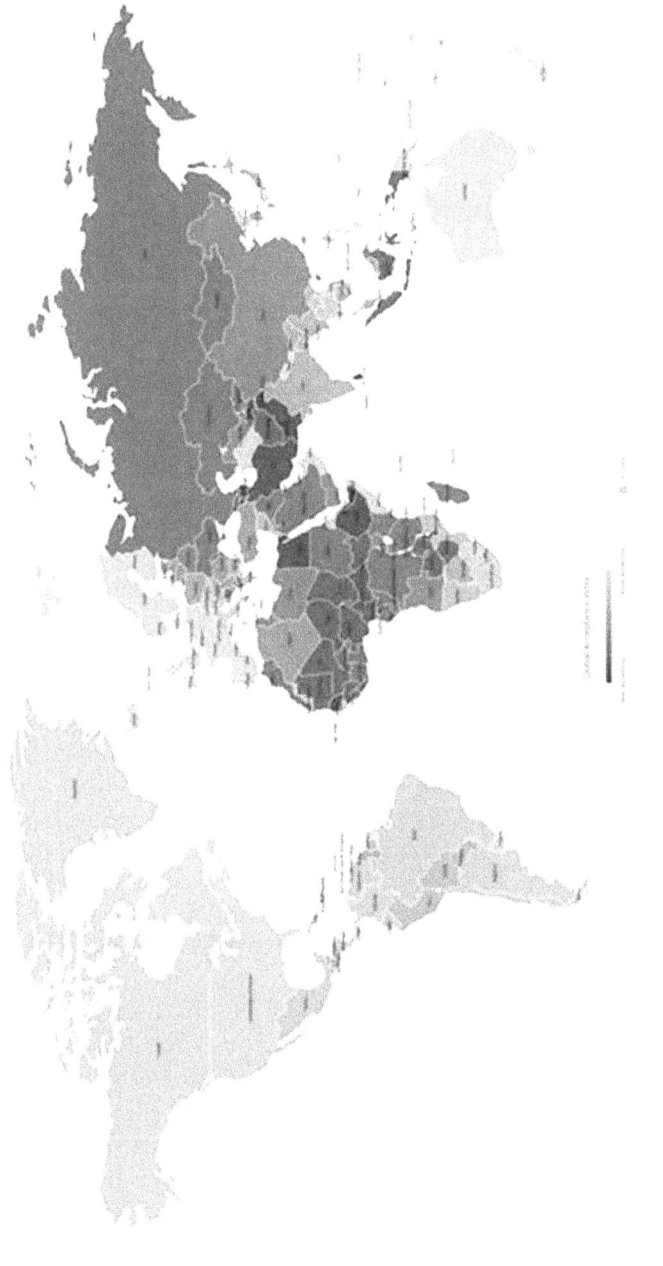

Figure 6.1 Global acceptance index (Source: William Institute, 2021)

The idea of those authors is that the generational shift represents a major force that will shape the future of tourism. That is to say that destinations and travel brands are seeking to cater to LGBTQ+ travellers facing their own challenges. But what does that mean for accommodations and destinations that offer and market vacations for young LGBTQ+? Some research results from this special issue show that although aspects such as convenience and good value for money are important to millennials and Gen Z: 'There is an important aspect of identification and a sense of belonging to the hostel community, which is increased when hostels offer experiences that appeal to the creation of emotional bonds. Therefore, it is not a coincidence that the best hostels are also the ones that enhance some emotional aspects, such as a family house with a meal cooked by "the mamma", or a hostel where they can have the possibility to make friends for life, with staff and other guests' (Veríssimo & Costa, 2018: 65).

Moreover, these 'new' generations travel more than any other generation, including baby boomers, and they are more likely to travel abroad than previous generations (Euromonitor International, 2019; ILGTA, 2019). The relation between young generations and tourism is attracting increasing attention also in order to gain a better understanding of tourism's future inclusiveness, to capture new trends of gender equality and to orient the concept of the gay destination. This is particularly important to counter the uncivil resistance of those countries where LGBTQ+ people are still victims of discrimination and social stigma and their quality of life has low levels due to the criminalisation of homosexual behaviours or because homosexual people cannot marry or same-sex couples cannot adopt. From this perspective, the rate of LGBTQ+ identification is increasing among the younger generations (GenForward, 2018).

Young generations of LGBTQ+ claim recognition of their personality on a social level, giving new meaning to sexual identities. Thus, they promote a sexuality that we could define as 'ductile', using Giddens' (1993) expression. Above all, millennials and Gen Zers are more self-aware, and, through their actions, they fight stereotypes based on gender and sexual orientation. Thus, from a tourist point of view, millennials can be considered as explorers, looking more for experiences than material things (Resonance Consultancy, 2018). These characteristics are also reflected in their tourism choices: they travel a lot, use the internet to make bookings, explore unusual destinations and avoid mass tourism to live enriching experiences from a human and cultural point of view (Canada Tourism Commission, 2015; Richards, 2007). They make tourist decisions considering also the social and environmental aspects, even if this involves spending extra money (Barton *et al.*, 2013). Ownership is no longer the priority of this age group. Unforgettable memories and compelling experiences are more valuable to them than material goods. These generations are willing to give up things like new clothes or cars so they can spend more money on travelling. Being digital natives, millennials and Gen Zers have a special relationship with technology. Plugged to their devices 24/7, they love using technology to make their life easier, especially when it comes to travelling. Some chains of boutique hotels embrace their young guests, letting them order from an emoji room service menu by texting a string of emojis (including their last name and room number) before being served by a robotic bellhop. In the same vein, Yotel, who claim to 'disrupt the world of hospitality through technology, design and our people', employ a robot that performs the activities of a bellboy carrying luggage before and after check-in.

Similarly, some studies have focused on LGBTQ-friendly destinations and possible constraints in relation to youths' travels (Szarycz, 2008; Weaver, 2011; Weeden et al., 2011). A recent Italian study (Bartoletti & Giannini, 2019) focuses on gay and lesbian young travellers. On the basis of 55 qualitative interviews of 29 gay men and 26 lesbian women aged 19–31 years, the study explored the relationship between sexuality, identity and travel, stressing the ambivalent perception of respondents regarding 'gay-friendly' tourist offers. For some interviewees, the existence of gay-friendly spaces and facilities in destinations is important. For other respondents these elements are not relevant. In particular, the single lesbian women interviewed show an attitude to live their tourist experiences regardless of their sexual orientation when a safe and welcoming environment is assured.

With regard to the issue of supportive and welcoming travel destinations, American research based on the focus group method (Smith, 2020) compares the homosexual and heterosexual attitudes of young generations. Sexual identity is an important aspect of a young person. Understanding and expressing sexual orientation and gender and developing related identities are important in the tourist mobility of youths. Many young LGBTQ+ people have trip motives that are no different from their heterosexual counterparts. Interests such as culture, languages, food, shopping, landscapes, good weather and historical sites are inviting to most people in general. What sets younger rainbow generations apart from others in their generation is the need for 'rainbow spaces'. Tourism destinations that already market to LGBTQ+ people are making sure that they identify facilities and products that could be considered of interest to young homosexuals. Having these different products can help in the creation of new forms of gay spaces and build a welcoming community that is essential to find in the destination for those who wish to experience it. When a homosexual person comes to a particular destination with an established LGBTQ+ community, they have a place to start. That being said, it is crucial to note that gay spaces or LGBTQ+-related entertainments are not for everyone in this segment; however, they can serve to invite them to the destination and encourage them to enjoy all that the touristic area offers.

Being one of the 'mega trends' that have significantly impacted the tourism system (Corbisiero & Paura, 2020), the role and use of social media in millennials' and Gen Z travellers' decision-making and in tourism operations and management have been widely discussed in tourism research. In the era of the Net Generation, the internet has evolved from a broadcasting medium to a participatory platform which allows young generations to become the 'media' themselves for collaborating and sharing information (Leung *et al.*, 2013; Li & Wang, 2011; Thevenot, 2007). The wide managing of social media provides LGBTQ+ youth with daily access to a broader sociocultural dialogue that shapes narrative identity development about trips and mobility in general (Monaco, 2019a). Social media are playing an increasingly important role as information sources for LGBTQ+ travellers. The internet has fundamentally reshaped the way tourism-related information is distributed and the way LGBTQ+ people plan for and consume travel (Buhalis & Law, 2008).

Compared to the old bestseller *Spartacus International Gay Guide* used above all by baby boomers to reach the few rainbow destinations around the world, the internet has fundamentally reshaped the rainbow mobility, underscoring changes that can significantly impact the tourism system. Social media websites, representing various forms of consumer-generated content such as blogs, virtual communities, social networks, tagging and media files shared on popular websites, have gained substantial popularity in online travellers' use of the internet (Gretzel, 2006; Pan *et al.*, 2007). Many of these social media websites assist consumers in posting and sharing their travel-related comments, opinions and personal experiences, which then serve as information for others. As such, search engines, for example, have become a powerful interface that serves as the 'gateway' to travel-related information as well as an important marketing channel to create LGBTQ+-friendly destinations as well as to spread 'pink tourism enterprises' and persuade potential visitors.

Tourism scholars agree that prioritising an organisation's website is an essential part of an excellent marketing strategy for young LGBTQ+ people, above all if the site has lots of welcoming and LGBTQ+-focused content as well as specific itineraries, bars, clubs and events geared towards that generation of travellers. Media partners such as paper and online magazines, publications and social media are also essential in getting the word out about LGBTQ+ travel information. It is, however, crucial to understand that the LGBTQ+ community is not the same throughout, and you cannot merely market to them as a whole.

Young LGBTQ+ individuals give top priority to inclusiveness, acceptance and safety. They want to visit places that have a good reputation for a favourable homosexual welcoming. Considering that the history of violence against LGBTQ+ people all over the world is made up of assaults on gay men, lesbians, bisexuals and transgender individuals and the current homophobic climate in some world areas, we could argue that their safety is at more of a risk, and therefore more of a concern. Fear for safety in the LGBTQ+ community is genuine because there is a history of unwelcomeness and even violence. The examples of Stonewall and the Upstairs Lounge Fire highlight some of the dark times. Although undoubtedly LGBTQ+ millennials and Gen Z are much safer today expressing themselves than in the past, these generations – shaped by the generations that came before them – are still fighting similar battles and brand new ones today. Even in modern times, massacres, like the Pulse nightclub shooting, still make safety a primary concern even in 2020. Millennials and Gen Z of all sexual orientations and gender identities are – as consumers and travellers – behaving differently from generations that have come before them. They seem less bothered by labels and less inclined to conform to rigid stereotypes, and much more accepting of LGBTQ+ people in general, raising questions over whether the segregation of homosexual travellers that has occurred in some destinations will be necessary or even desirable in the future (UNWTO, 2017).

It is well known that LGBTQ+ people are still the victims of verbal and even physical abuse in any setting. It is crucial to understand that even non-offensive language or behaviour can be offensive to homosexual people in tourism.

Poria (2006) gave an example of how many of the negative experiences that LGBTQ+ people had while travelling came from hotel staff. An examination of the hotel experiences of lesbians and gay persons in the UK and Israel showed that homosexual tourists are highly affected by the symbolic meanings they assign to specific elements of the hotel experience. The results of this research indicate that it is important for gay men and lesbians to feel accepted and welcome when their sexual orientation is known.

Conclusion

By integrating the sociological lens with the generational analysis of tourism, this chapter has charted the trajectory of the future of tourism by adding a further element of discussion: LGBTQ+ travellers. We began by arguing that although the future of tourism can be better understood through the lens of youth homosexual travellers, the road ahead for LGBTQ+-sensitive tourist hospitality still presents obstacles.

The findings of tourism sociological research may assist in leading both service leisure provision and hospitality for homosexual people. It is especially important for local communities and destinations that want to be welcoming to LGBTQ+ people, to have some form of diversity training in public places such as hotels in order to ensure that everyone is treated with due respect.

Fortunately, in recent years many touristic countries have taken effective measures to combat discrimination. For decades, travelling provided an escape from restrictive legislations and oppressive social situations to countries where homosexual activity was more easily accessed and enjoyed (Waitt & Markwell, 2006). The increased significance of rainbow tourism occurred in parallel with a sense of 'intersectional' tourist hospitality that does not discriminate between minorities but, on the contrary, would understand and accept their specificities. The gay tourism industry and the IGLA organisation have been major players in this process along with gay (and not gay) guidebooks and, of course, the internet and social networks. They helped several countries to make this change that also includes removing criminal sanctions for same-sex behaviour, legal prohibition of discrimination on the basis of sexual orientation, gender identity and intersex status, legal recognition of the gender identity of transgender persons without abusive requirements, legal recognition of same-sex couples and their families or caregivers, public education and awareness raising campaigns to combat homophobia and transphobia, establishing shelters for homosexual homeless and youth, and anti-bullying initiatives in schools. Rainbow destinations have also been pioneers in combatting homophobia and transphobia, in turn transforming the debate at the national level in their respective countries and sustaining tourist destinations as spaces of liberation and democracy. In doing so, the apparent sense of freedom is in part a paradox because of the sanctification, masculinisation and sexualisation of gay destinations that restricted them to a specific socially acceptable expression of 'gayness' and 'whiteness'. The empirical research suggested that several destinations have had a post-colonial attitude in LGBTQ+ hospitality. For instance, Waitt's (2003) field work showed how some gay men living in Sydney did not take part in the Sydney 2002 Gay Games because of either the costs of registration or the 'macho' construction of gayness employed in the branding strategies.

However, this rapidly growing segment is expected to become more and more visible of all travellers by 2030 (IGLTA, 2020).

World exploration, social interaction, emotional experience, sense of community: many brands in the hotel industry have realised that they need to rethink the service they provide to accommodate these dimensions. Millennials expect a greater link between tourism services and their everyday life and as a result many new hotel brands and leisure organisations are arising which try to provide homosexual travellers with precisely this experience. The entire hospitality chain is being adjusted to meet their lifestyle requirements and will continue to do so while becoming more welcoming, LGBTQ+ friendly and above all tech-savvy, with a strong focus on empathy and rainbow traveller connection. The trend is expected to be further boosted by the accelerated implementation of digitalisation, as technology is essential for this target, especially for the youngest. As we argued in the previous chapters, NetGen, iGen or 'click n go' generations (Benckendorff *et al.*, 2010) represent quite a radical shift from baby boomers and the oldest generations. As for future strategic planning, there is no real need to prepare any 'special offers' for LGBTQ+ travellers. This target would need first of all the sense of security and expression of tolerance which are what most matter to them. Destinations, tourist companies and the hospitality sector which are responsible for the creation and development of competitive tourist destinations should play an important role in the post-Covid future, by planning appropriate promotional actions which can help create the image of a given destination as being LGBTQ+ friendly.

Conclusion

Understanding the Present to Prepare for the Future (of Tourism)

Fabio Corbisiero, Salvatore Monaco and Elisabetta Ruspini

The analysis proposed in the previous pages has made it clear that nowadays tourism is experiencing a season of change. The factors that are defining a new way of understanding and experiencing tourist choices concern both aspects of a structural nature (e.g. systems of values, socio-technological devices, means of transport, alternative forms of economy, territorial transformations), and the forms of sociality that take place at the micro level.

In this scenario, the use of the generational perspective to understand both the characteristics of tourism in the contemporary world and its possible developments in the future has been a very useful tool. In fact, through the focus on the younger generations we have been able to intercept some of the main directions that are giving new shape and substance to tourist mobility.

The sociological portrait of the new generations that has been proposed in this volume allows us to argue that, on a global level, today's young people can be considered agents of change. In fact, members of new generations have visions of the world and of life that are very different from those of older people. More specifically, the main characteristics of younger people that our analysis made it possible to identify are: their perception of closer distances and the easier reachability of territories and destinations that the older generations used to see as far and sometimes difficult to get to; their competency to obtain information and images in real time; their concern about environmental issues and the care of the world; their activism for equal rights and against gender inequalities and social differences.

These visions, values and ideals obviously also flow into their tourism choices, as well as their flexibility and ability to cope with changes that occur at various dimensions of daily life. It is important to remember that the younger generations have both experienced economic, social and environmental crises and made a significant contribution to overcome them by applying creative and innovative solutions. The generational lens makes it possible to argue that their resilience and openness to change will help them face the new challenges that the pandemic has created, also in the tourism field.

For its extraordinary nature, the Covid-19 pandemic crisis can be considered a turning point. It has overwhelmed and upset the whole world, marking a clear break with the past we were accustomed to. Indeed, the rapid global spread of the coronavirus has forced all people to redefine their daily lives (e.g. Abulibdeh, 2020; Carpenter & Dunn, 2020; Di Nicola & Ruspini, 2020; Jenkins & Smith, 2021; Kenway & Epstein, 2021; Monaco & Nothdurfter, 2021; Trinidad, 2021; Yates *et al.*, 2020; Žižek, 2020). The 'new normal' (Jesus *et al.*, 2020) has affected all areas of associated life, from education to work, from family relationships to leisure activities, inevitably also passing through tourism. Regarding this specific aspect, it is safe to argue that tourism in the pandemic era has been characterised by new rules and, at the same time, it has generated new needs in travellers. 'Making tourism safe' can be considered a new paradigm (Monaco, 2021), since people have not lost their desire to explore the world, but they must find the most suitable strategies to mediate between their will to travel and the need for safety. Nowadays, travellers divert their attention towards tourist destinations characterised by maximum safety, protecting themselves from the risks associated with the pandemic. In this sense, all operators in the tourism supply chain must commit themselves not only to offering quality services and products, but also to instilling a sense of security, adapting to the main health protection protocols currently in force, also with the help of new technologies.

Based on our analysis, it is possible to argue that younger people are already adapting to the 'covidised' world (Pai, 2020) and to the new travel rules.

A first important consideration to be made concerns the role of new technologies in the fight against Covid-19. As we know, digital technologies have played (and still are playing) a key role in addressing the effects of the pandemic. Not surprisingly, the world underwent an important digitisation process in 2020 (European Commission, 2020). Not only were new websites, e-services and online shops born, but also apps to make reservations or check one's own health status (e.g. Akiyama, 2020; Lee & Lee, 2020; Watanabe & Omori, 2020). Similarly, some technological innovations have proved to be excellent allies in monitoring and limiting the trend of infections, albeit with different results depending on the country. We can think, for example, of infection tracking apps or scanners capable of automatically detecting people's body temperature which quickly spread all over the world (e.g. Hendl *et al.*, 2020; Walrave *et al.*, 2020; Wang *et al.*, 2020).

As it is known, already in the pre-Covid-19 period one of the main features of millennials and Generation Z was their possession and ability to use mobile devices (such as smartphones and tablets). Consequently, they have immediately learned how to manage the most appropriate technological solutions to orient themselves in the new daily life, disrupted by the health crisis, even in the tourism sector. Through their devices connected 24 hours a day, members of younger generations became familiar with maps updated in real time to identify the destinations most affected by the virus, government sites for knowing travel restrictions, quick response (QR) codes to access digital content of interest to them through their mobile phones (such as menus or catalogues) and new user experience services to conclude economic transactions or to communicate with actors in the tourism sector.

Secondly, several studies on the subject (e.g. Das & Tiwari, 2020; Hussain & Fusté-Forné, 2021; Woyo, 2021) have highlighted that the global crisis has given a new impetus to local and domestic tourism. Indeed, as international mobility has come to a halt, people have begun to choose as tourist destinations some places closer to their homes. This phenomenon is partly explained by the desire of many people to have the peace of mind of being able to return to their city in the event of sudden problems caused by Covid-19, such as the identification of new outbreaks.

This type of attitude, which for some people represented a temporary fallback, seems to have been a very welcome solution for the younger generations. In fact, from the millennial generation onwards, great attention has been paid to safeguarding the environment, which in many cases has also resulted in the search for tourism solutions with a low or limited environmental impact. Short-term trips fit perfectly within this framework, effectively representing a solution capable of providing more sustainable tourist experiences.

Clearly, the coronavirus has posed several challenges other than environmental issues, which primarily concern people's health and wellbeing. However, the pandemic has implicitly contributed to highlighting the interconnected nature of social systems and environmental contexts. More specifically, it is possible to argue not only that the health and the climate emergencies have some points in common, but also that they mutually influenced each other. Actually, both emergencies share the fact that they have significant implications at the local, national and global level, and, consequently, both require the identification of solutions to be activated locally, nationally and globally (Lord, 2021). At the same time, the immobility imposed by the lockdowns and the closure of the borders has discouraged the use of the most polluting means of transport, favouring outdoor walks and the use of more sustainable means of transport, such as bicycles and scooters (Corbisiero & La Rocca, 2020). In other words, the impossibility of leaving their own regions has also represented a solution to reduce the consumption of fossil fuels, encouraging on a large scale the spread of a more sustainable and authentic tourism (Perkins *et al.*, 2021).

This revolution in tourist habits has had an interesting impact on the environment and the planet (EEA, 2021). One of the most extraordinary short-term effects has been the reduction of greenhouse gas emissions globally. For example, in the course of 2020, in regard to Europe only, a reduction in greenhouse gas emissions of 7.6% was recorded. The reasons behind this milestone are directly linked not only to a major change in work and living habits (with work from home and the reduction of business and tourist travels), but also by the fact that pandemic inspired locals to rediscover natural wonders. As a consequence, the entire transport industry has seen a decline in the use of cars, ships, airplanes and trains, and consequently, a slump in emissions.

Another positive indicator that testifies to the environmental improvement is represented by the reduction of atmospheric pollution: in fact, the reduction in transport has lowered the concentrations of NO_2, PM_{10}, $PM_{2.5}$ and other pollutants which are the most harmful elements that worsen the air quality. In the most polluted cities, such as Milan and Madrid, this reduction has even reached 70% (Collivignarelli *et al.*, 2020; Donzelli *et al.*, 2020; Nigam *et al.*, 2020; Rodríguez-Urrego & Rodríguez-Urrego, 2020).

In the main cities of the world, equally positive short-term impacts have been observed also on noise pollution levels, which were certainly affected by traffic restrictions and reductions in air transport traffic, and other noisy activities. The blockade period also showed how animal and plant species react to less human disturbance, both in rural and urban contexts. The smaller number of tourists in circulation – albeit for a rather limited period – has offered ecosystems and habitats the opportunity to regenerate, occupying new spaces and territorial niches. Learning from this lesson, global cities, which in the past have often been the protagonists of overtourism, should increase the number of parks and green spaces within them to promote better coexistence among different species.

From an analytical point of view, it is possible to argue that unintentionally the new scenario that resulted from the pandemic can help the so-called 'Greta Generation' to spread its vision of the world, also urging other generations (past and future ones) to engage in a fight for a just transition, without the distinction of gender, ethnicity, sexual identity or religion to promote a greener and more sustainable future.

One of the most important lessons that can be learned from the pandemic is linked to the possibility of adopting new and more sustainable lifestyles and tourist habits. The temporary suspension of mass tourism has shown how the bad habits that characterised the pre-Covid tourism market are capable of negatively impacting the fate of territories and people (e.g. Milano & Koens, 2021; Miller, 2021). It is essential to help and include marginalised and vulnerable groups in recovery efforts, while focusing on environmental conservation to help the planet thrive. The 'new normal' that is being created must provide for both people and planet, requiring a holistic and collective approach to sustainability.

Members of the younger generations, online and through strikes and demonstrations, had long ago stressed the need to rethink, re-evaluate and reorganise the tourist flows, favouring trips with a low environmental impact, more sustainability and respect for plant and animal species that inhabit the destinations visited. In this sense, the pandemic has further strengthened the desire of younger generations to contribute to a better world in which societies and national governments are committed to generating a positive social impact, putting people and environmental sustainability ahead of profit.

According with our findings, younger generations have demonstrated a profound resilience in the face of the new challenges imposed by the pandemic (Deloitte, 2021). On the one hand, young people are concerned about the economic, financial and social impact of Covid-19, but on the other hand, the majority of them think that the pandemic crisis may represent an opportunity to start on new foundations and build a more equitable, sustainable and inclusive society. Globally, millennials and Gen Zers hope a better world will emerge from the pandemic, and young people want to lead this change. Thus, they could, in particular, take advantage of the post-pandemic period of social and cultural reorganisation, acting as a spokesperson for the importance of enhancing natural areas and redeveloping urban environments. The inclusion of young people in policymaking will thus ensure that recovery is sustainable and inclusive.

In summary, the crisis that the tourism sector experienced between 2020 and 2021 due to mobility restrictions could represent the opportunity for the new generations to make more concrete the battles they were already fighting, insisting on the possibility of aiming for a green tourist restart.

In this scenario, the deep socioeconomic crisis that has affected the energy markets can provide incentives for a more environmentally sustainable global energy transition.

Among the most important factors that are necessary for these generational goals to be disseminated, we can find: the collaboration of science, the intervention of public policy, and international cooperation. In fact, the Covid-19 pandemic has made it clear that collaboration among sectors in different territories is a very powerful lever to tackle urgent global issues. To address future challenges, such as climate change and the need to reorganise tourism, in line with the expectations and needs of the new generations, governments and science should be able to open a constructive and bilateral dialogue with representatives of the new generations in a coordinated manner, providing them with concrete and more consistent support.

The restrictions imposed on mobility to stem the pandemic have increased the interest in domestic and proximity tourism: there has been and will probably continue to be a net reduction in carbon dioxide and other greenhouse gas emissions. Understandably, during the most acute phase of the crisis, governments focused primarily on providing short-term financial assistance and economic support to sectors primarily affected by the crisis to help them recover as quickly as possible. In the long term, the new effort to be made should be towards including elements capable of shifting the economy and tourism in a green and climate-friendly direction, characterised by less dependence on fossil fuels, greater dependence on renewable energy sources and greater efficiency in the production and use of energy.

Looking at the future, it is possible to argue that the 'new normal' we are getting used to will represent something ordinary for future generations. Just think that Generation Alpha children are still in a phase of personal training and growth. Thus, even if they have had some opportunities to travel, because of their very young age they have certainly not yet experienced the possibility of living the travel experience in absolute autonomy. Consequently, they will naturally learn all the new rules and procedures for travelling that are now mandatory for safety reasons: wearing masks, having their temperature measured and keeping a safe distance are just some of the social norms that the younger generations will probably assimilate without asking too many questions in future. Thus, it is not too much to assume that even when the health emergency is completely over, they will continue to engage in some of the behaviours they are learning today, since, for them, these are the normal rules to follow when travelling.

Finally, the difficult sociopolitical situation experienced by younger generations has further worsened with the pandemic crisis. Covid-19 has posed severe challenges to young people: disrupted family routines and relationships, family stress, learning and job loss, increased fear and anxiety. The impact of the pandemic on young people is likely to be intense and long lasting. To avoid exacerbating intergenerational inequalities, young women and men need to be fully included in the international sociopolitical agenda. A crucial element in successfully designing the post-pandemic society is to understand the values and lifestyles of millennials and Gen Zers, and the specificities that chacterize women, men and LGBT individuals, also in order to predict their future needs and support their contribution to the ongoing discussions on sustainable recovery.

References

Aboim, S. and Vasconcelos, P. (2013) From political to social generations: A critical reappraisal of Mannheim's classical approach. *European Journal of Social Theory* 17 (2), 165–183.
Abulibdeh, A. (2020) Can Covid-19 mitigation measures promote telework practices? *Journal of Labor and Society* 23 (4), 551–576.
Accenture (2018) *Global Consumer Survey 2018*. New York: Accenture.
Adamy, J. (2020) Millennials slammed by second financial crisis fall even further behind. *The Wall Street Journal*, 9 August.
Akhavan Sarraf, A.R. (2019) Generational groups in different countries. *International Journal of Social Sciences and Humanities* 4 (1), 41–52.
Akiyama, I. (2020) Basic and recent applied technologies of ultrasound in the field of clinical diagnosis. *J-Stage* 41 (6), 845–850.
Aldrich, R. (1993) *The Seduction of Mediterranean: Writing, Art and Homosexual Fantasy*. London: Routledge.
Aldrich, R. (2004) Homosexuality and the city: An historical overview. *Urban Studies* 41, 1719–1737.
Algar, R. (2007) Collaborative consumption. *Leisure Report* 4, 16–17.
Allen, R.S., Allen, D.E., Karl, K. and White, C.S. (2015) Are millennials really an entitled generation? An investigation into generational equity sensitivity differences. *Journal of Business Diversity* 15 (2), 14–26.
Alwin, D. and McCammon, R. (2007) Rethinking generations. *Research in Human Development* 4 (3–4), 219–237.
Anderson, M. and Jiang, J. (2018) Teens, social media and technology 2018. Pew Research Center, 31 May. See https://www.pewresearch.org/internet/2018/05/31/teens-social-media-technology-2018/ (accessed 8 October 2020).
Andrejevic, M. and Burdon, M. (2014) Defining the sensor society. *Television and New Media* 16 (1), 19–36.
Appadurai, A. (1993) Disjuncture and difference in the global cultural economy. In M. Featherstone (ed.) *Global Culture, Nationalism, Globalization and Modernity* (pp. 295–310). London: Sage.
Appadurai, A. (1996) *Modernity at Large: Cultural Dimensions of Globalization*. Minneapolis, MN: University of Minnesota Press.
Appiah, K.A. (2007) *The Ethics of Identity*. Princeton, NJ: Princeton University Press.
Ariès, P. (1979) 'Generazioni'. In AA. VV (eds) *Enciclopedia Einaudi*. Turin: Einaudi.
Artal-Tur, A. and Kozak, M. (2019) *Culture and Cultures in Tourism: Exploring New Trends*. New York: Routledge.
Atzori, L., Iera, A. and Morabito, G. (2010) The internet of things: A survey. *Computer Networks* 54 (15), 2787–2805.

Backman, K.F., Backman, S.J. and Silverberg, K.E. (1999) An investigation into the psychographics of senior nature-based travellers. *Tourism Recreation Research* 24 (1), 13–22.

Bain and Company (2019) *Automotive and Mobility Insights*. Mexico City: Bain and Company Press.

Barbosa, B. and Fonseca, I. (2019) A phenomenological approach to the collaborative consumer. *Journal of Consumer Marketing* 36 (6), 705–714.

Barroso, A., Parker, K. and Bennett, J. (2020) As millennials near 40, they're approaching family life differently than previous generations: Three-in-ten millennials live with a spouse and child compared with 40% of Gen Xers at a comparable age. Pew Research Center, Washington CD. 27 May. See https://www.pewresearch.org/social-trends/2020/05/27/as-millennials-near-40-theyre-approaching-family-life-differently-than-previous-generations/ (accessed 18 November 2021).

Bartoletti, R. (2001) L'innovazione nell'industria culturale globale, tra globalizzazione e indigenizzazione. *Studi di Sociologia* 39 (2), 147–161.

Bartoletti, R. and Giannini, L. (2019) Perché devo dire qual è il mio orientamento sessuale se voglio farmi semplicemente una vacanza? *Fuori Luogo. Rivista di Sociologia del Territorio, Turismo, Tecnologia* 5 (1), 8–18.

Barton, C., Haywood, J., Jhunjhunwala, P. and Bhatia, V. (2013) Traveling with millennials, BCG-Boston Consulting Group. See https://image-src.bcg.com/Images/BCG%20Traveling%20with%20Millennials%20Mar%202013_tcm9-96752.pdf (accessed 27 August 2020).

Bauman, Z. (1998) *Globalization: The Human Consequences*. Cambridge: Polity Press.

Bec, A., Moylea, B., Timms, K., Schaffer, V., Skavronskaya, L. and Little, C. (2019) Little management of immersive heritage tourism experiences: A conceptual model. *Tourism Management* 72, 117–120.

Beccalossi, C. (2015) *Italian Sexualities Uncovered, 1789–1914. Genders and Sexualities in History*. London: Palgrave MacMillan.

Beck, U. and Beck-Gernsheim, E. (2001) *Individualization: Institutionalized Individualism and Its Social and Political Consequences*. London: Sage Publications.

Becker, H.P. (1956) *Man in Reciprocity: Introductory Lectures on Culture, Society, and Personality*. Westport, CT: Greenwood Press.

Becker, H.A. (1992a) *Dynamics of Cohort and Generations Research*. Amsterdam: Thesis.

Becker, H.A. (1992b) *Generations and Their Opportunities*. Amsterdam: Meulenhof.

Belch, G.E. and Belch, E. (2015) *Advertising and Promotion. An Integrated Marketing Communication Perspective*. New York: McGraw-Hill.

Beldona, S., Nusair, K. and Demicco, F. (2009) Online travel purchase behaviour of generational cohorts: A longitudinal study. *Journal of Hospitality Marketing and Management* 18 (4), 406–420.

Bell, W. (2009) *Foundations of Futures Studies, Volume 1: Human Science for a New Era*. New Brunswick, NJ: Transaction Publishers.

Belsky, E.S., Herbert, C.E. and Molinsky, J.H. (eds) (2014) *Homeownership Built to Last: Balancing Access, Affordability, and Risk after the Housing Crisis*. Washington, DC: Brookings Institution Press.

Benckendorff, P., Moscardo, G. and Pendergast, D. (eds) (2010) *Tourism and Generation Y*. Wallingford: CABI Publishing.

Benckendorff, P.G., Xiang, Z. and Sheldon, P.J. (2019) *Tourism Information Technology* (3rd edn). Boston, MA: CABI.

Berkhout, P., Achterbosch, T., Van Berkum, S., Dagevos, H., Dengerink, J., Van Duijn, A.P. and Terluin, I.J. (2018) Global implications of the European food system: A food systems approach. *Wageningen Economic Research* 51.

Berkup, S.B. (2014) Working with generations X and Y in generation Z period: Management of different generations in business life. *Mediterranean Journal of Social Sciences* 5 (19), 218–229.

Bernardi, M. and Ruspini, E. (2018) 'Sharing tourism economy' among millennials in South Korea. In Y. Wang, A. Shakeela, A. Kwek and C. Khoo-Lattimore (eds) *Managing Asian Destinations* (pp. 177–196). Singapore: Springer.

Beutell, N. and Wittig-Berman, U. (2008) Work–family conflict and work–family synergy for generation X, baby boomers, and matures: Generational differences, predictors, and satisfaction outcomes. *Journal of Managerial Psychology* 23, 507–523.

Bialeschki, M.D. (2005) Fear of violence: Contested constraints by women in outdoor recreation activities. In E.L. Jackson (ed.) *Constraints to Leisure* (Chapter 7). State College, PA: Venture Publishing.

Bialik, K. and Fry, R. (2019) Millennial life: How young adulthood today compares with prior generations. Pew Research Center Social and Demographic Trends. 14 February. See https://www.pewsocialtrends.org/essay/millennial-life-how-young-adulthood-today-compares-with-prior-generations/ (accessed 10 August 2020).

Biella, A. and Borzini, G. (2004) *L'evoluzione del sistema agenziale verso la vendita online*. Milan: FrancoAngeli.

Bigné, E., Andreu, L., Hernandez, B. and Ruiz, C. (2016) The impact of social media and offline influences on consumer behaviour. *Current Issues in Tourism* 1 (19), 291–313.

Blázquez, A. (2021) *Consumers Who are Vegan or Vegetarian in the U.S. 2018, by Age Group*. New York: Statista.

Blichfeldt, B., Chor, J. and Milan, N. (2011) It really depends on whether you are in a relationship: A study of gay destinations from a tourist perspective. *Tourism Today* 3, 7–26.

Bocken, N., Short, S., Rana, P. and Evans, S. (2014) A literature and practice review to develop sustainable business model archetypes. *Journal of Cleaner Production* 65, 42–56.

Bolton, R.N., Parasuraman, A., Hoefnagels, A., Migchels, N., Kabadayi, S., Gruber, T., Komarova Loureiro, Y. and Solnet, D. (2013) Understanding generation Y and their use of social media: A review and research agenda. *Journal of Service Management* 24 (3), 245–267.

Bolzendahl, C.I. and Myers, D.J. (2004) Feminist attitudes and support for gender equality: Opinion change in women and men, 1974–1998. *Social Forces* 83 (2), 759–789.

Bond, M. (1997) *Women Travellers: A New Growth Market*. Pacific Asia Travel Association Occasional Paper 20.

Bontekoning, A.C. (2011) The evolutionary power of new generations: Generations as key players in the evolution of social systems. *Psychology Research* 1 (4), 287–301.

Bontekoning, A.C. (2018) *The Power of Generations*. Amsterdam: Warden Press.

Book, L.A., Tanford, S., Montgomery, R. and Love, C. (2018) Online traveller reviews as social influence: Price is no longer king. *Journal of Hospitality & Tourism Research* 42 (3), 445–475.

Booking (2019b) Gen Z and the future of sustainable travel. See https://globalnews.booking.com/gen-z-and-the-future-of-sustainable-travel/ (accessed 10 August 2020).

Booking (2020) *Booking.com's 2019 Sustainable Travel Report*. Mumbai: Booking.com.

Botsman, R. (2017) *Who Can You Trust? How Technology Brought Us Together and Why It Might Drive Us Apart*. London: PublicAffairs.

Bourdieu, P. (1980) *Le Sens Pratique*. Paris: Minuit.

Bourdieu, P. (1990) 'Youth' is just a word. In P. Bourdieu (ed.) *Sociology in Question* (pp. 94–102). Thousand Oaks, CA: Sage.

Bourdieu, P. (1993) *The Field of Cultural Production: Essays on Art and Literature*. Cambridge: Polity Press.

Bourguinat, N. (2016) Women's travels in Europe: 19th–20th centuries. In *Encyclopédie pour une histoire nouvelle de l'Europe* [online], published 10 November 2017. See https://ehne.fr/en/article/gender-and-europe/european-circulations-shifting-gender/womens-travels-europe (accessed 3 March 2020).

Bouton, C.W. (1965) John Stuart Mill: On liberty and history. *The Western Political Quarterly* 18 (3), 569–578.

Boyd, D. and Ellison, N.B. (2007) Social network sites: Definition, history, and scholarship. *Journal of Computer-Mediated Communication* 13 (2), 210–230.

Brewster, K.L. and Padavic, I. (2000) Change in gender-ideology, 1977–1996: The contributions of intracohort change and population turn-over. *Journal of Marriage and Family* 62 (2), 477–487.

Briatte, A-L. (2016) Feminisms and feminist movements in Europe: XIX–XXI. In *Encyclopédie pour une histoire nouvelle de l'Europe* [online], published 13 November 2019. See https://ehne.fr/en/article/gender-and-europe/feminisms-and-feminist-movements/feminisms-and-feminist-movements-europe (accessed 31 March 2020).

Bristow, J. (2015a) Understanding generations historically. In J. Bristow (ed.) *Baby Boomers and Generational Conflict* (pp. 19–41). Basingstoke: Palgrave MacMillan.

Bristow, J. (2015b) Mannheim's 'problem of generations' revisited. In J. Bristow (ed.) *Baby Boomers and Generational Conflict* (Chapter 3). Basingstoke: Palgrave MacMillan.

Bristow, J. (2015c) The boomers as an economic problem. In J. Bristow (ed.) *Baby Boomers and Generational Conflict* (Chapter 6). Basingstoke: Palgrave MacMillan.

British Airways (2018) (Don't) come fly with me. Press release 10 October 2018. See https://mediacentre.britishairways.com/pressrelease/details/86/2018-247/10174 (accessed 6 April 2020).

British Encyclopaedia (2019) *Instagrammabilty*. Oxford: Oxford University Press.

Brokaw, T. (2004) *The Greatest Generation*. New York: Random House.

Brooks, C. and Bolzendahl, C. (2004) The transformation of US gender role attitudes: Cohort replacement, social-structural change, and ideological learning. *Social Science Research* 33 (1), 106–133.

Brosdahl, D.J.C. and Carpenter, J.M. (2011) Shopping orientations of US males: A generational cohort comparison. *Journal of Retail and Consumer Service* 18 (6), 548–554.

Buhalis, D. and Law, R. (2008) Progress in information technology and tourism management: 20 years on and 10 years after the internet – the state of eTourism research. *Tourism Management* 29 (4), 609–623.

Butler, K.L. (1995) Independence for western women through tourism. *Annals of Tourism Research* 22 (2), 487–489.

Buzza, J.S. (2017) Are you living to work or working to live? What millennials want in the workplace. *Journal of Human Resources Management and Labor Studies* 5 (2), 15–20.

Canada Tourism Commission (2015) *Special Examination Report*. Grosse Île: Canada Tourism Commission.

Cantelmi, T. (2013) *Tecnoliquidità. La psicologia ai tempi di internet: la mente tecnoliquida*. Milan: San Paolo Edizioni.

Carpenter, D. and Dunn, J. (2020) We're all teachers now: Remote learning during Covid-19. *Journal of School Choice* 14 (4), 567–594.

Carty, M. (2019) Millennial and Gen Z traveller survey 2019: A multi-country comparison report. *Skift Research*. See https://research.skift.com/wp-content/uploads/2019/08/MillennialGenZ_Final.8.12.pdf (accessed 5 August 2020).

Casey, M. (2009) Tourist gay(ze) or transnational sex: Australian gay men's holiday desires. *Leisure Studies* 28 (2), 157–173.

Castells, M. (1996) *The Rise of the Network Society*. Malden, MA: Blackwell Publishers.

Castells, M. (2005) Global governance and global politics. *PS: Political Science & Politics* 38 (1), 9–16.

Castells, M. (2007) Communication, power and counter-power in the network society. *International Journal of Communication* 1, 238–266.

Cavagnaro, E., Staffieri, S. and Postma, A. (2018) Understanding millennials' tourism experience: Values and meaning to travel as a key for identifying target clusters for youth (sustainable) tourism. *Journal of Tourism Futures* 4 (1), 31–42.

CB Isights (2019) *Fintech Trends to Watch in 2019*. New York: CB Isights Research.

CBI (2019) *Tech Tracker 2019: The Must-Know Technology and Innovation Trends*. New York: CBI.

Chambers, D., Munar, A.M., Khoo-Lattimore, C. and Biran, A. (2017) Interrogating gender and the tourism academy through epistemological lens. *Anatolia* 28 (4), 501–513.

Chan, N.D. and Shaheen, S.A. (2012) Ridesharing in North America: Past, present, and future. *Wayback Machine Transport Reviews* 32 (1), 93–112.

Chen, W.C., Battestini, A., Gelfand, N. and Setlur, V. (2009) Visual summaries of popular landmarks from community photo collections. In *ACM Multimedia Conference*. Beijing: Association for Computing Machinery.

Chhetri, P., Hossain, M.I. and Broom, A. (2014) Examining the generational differences in consumption patterns in South East Queensland. *City, Culture and Society* 5 (4), 1–9.

Chiang, C.Y. and Jogaratnam, G. (2006) Why do women travel solo for purposes of leisure? *Journal of Vacation Marketing* 12 (1), 59–70.

Chiang, L., Manthiou, A., Tang, L., Shin, J. and Morrison, A. (2014) A comparative study of generational preferences for trip-planning resources: A case study of international tourists to Shanghai. *Journal of Quality Assurance in Hospitality and Tourism* 15 (1), 78–99.

Choe, Y. and Fesenmaier, D.R. (2017) The quantified traveller: Implications for smart tourism development. In Z. Xiang and D. Fesenmaier (eds) *Analytics in Smart Tourism Design. Tourism on the Verge* (pp. 65–80). Cham: Springer.

Chung, N., Han, H. and Joun, Y. (2015) Tourists' intention to visit a destination: The role of augmented reality (AR) application for a heritage site. *Computers in Human Behaviour* 50, 588–599.

Clarke, I.F. (1988) The right connections. *Tourism Management* 9 (1), 78–82.

Clift, S. and Wilkins, J. (1995) Travel, sexual behaviour and gay men. In P. Aggleton, P. Davies and G. Hart (eds) *AIDS: Safety, Sexuality and Risk* (pp. 35–54). London: Taylor & Francis.

CMI (2014) *CMI's 19th Annual LGBT Tourism and Hospitality Survey December 2014*. Corte Madera: Community Marketing and Insights.

CMI (2019) *CMI's 24th Annual Survey on LGBT Tourism and Hospitality: US Overview Report*. Corte Madera: Community Marketing and Insights.

Cohen, M. (2001) The grand tour: Language, national identity and masculinity. *Changing English: Studies in Culture and Education* 8 (2), 129–141.

Cohen, S.A. and Cohen, E. (2019) New directions in the sociology of tourism. *Current Issues in Tourism* 22 (2), 153–172.

Collins English Dictionary (2014) 'Gaycation'. In *Collins English Dictionary – Complete and Unabridged* (12th edn). London: HarperCollins Publishers.

Collins, D. and Tisdell, C. (2002) Gender and differences in travel life cycles. *Journal of Travel Research* 41, 133–143.

Collivignarelli, M.C., Abbà, A., Bertanza, G., Pedrazzani, R., Ricciardi, P. and Carnevale Miino, M. (2020) Lockdown for CoViD-2019 in Milan: What are the effects on air quality? *The Science of the Total Environment* 732, 139280.

Colombo, F. (2005) La ricerca sulla comunicazione tra locale e globale: territorio e virtualità. *Sociologia della comunicazione* 37, 85–96.

Comte, A. (1849) *Cours de philosophie positive* (Paris, 1849), IV, 635–641.

Cone Communications (2016) 2016 millennial employees engagement study. See https://static1.squarespace.com/static/56b4a7472b8dde3df5b7013f/t/5819e8b303596e3016ca0d9c/1478092981243/2016+Cone+Communications+Millennial+Employee+Engagement+Study_Press+Release+and+Fact+Sheet.pdf (accessed 18 March 2020).

Coomes, M.D. and DeBard, R. (2004) A generational approach to understanding students. *New Directions for Student Service* 4 (106), 5–16.

Cooper, D., Holmes, K., Pforr, C. and Shanka, T. (2019) Implications of generational change: European river cruises and the emerging Gen X market. *Journal of Vacation Marketing* 25 (4), 418–431.

Corbisiero, F. (2014) Homosexing... in the city: LGBT communities and rainbow tourism. In AA. VV (ed.) *Gender-Based Violence: Homophobia and Transphobia* (pp. 104–113). Milan: McGraw-Hill.
Corbisiero, F. (2016) *Sociologia del turismo LGBT*. Milan: FrancoAngeli.
Corbisiero, F. (2020) Sostenere il turismo: come il Covid-19 influenzerà il viaggio del future. *Fuori Luogo. Rivista di Sociologia del Territorio, Turismo, Tecnologia* 7 (1), 69–79.
Corbisiero, F. and Monaco, S. (2017) *Città arcobaleno. Una mappa della vita omosessuale in Italia*. Rome: Donzelli.
Corbisiero, F. and Ruspini, E. (eds) (2018) Millennials and generation Z: Challenges and future perspectives for international tourism. *The Journal of Tourism Futures-ETFI* (Special Issue) 4 (1), 3–6.
Corbisiero, F. and La Rocca, R.A. (2020) Tourism on demand: A new form of urban and social demand of use after the pandemic event. *Tema. Journal of Land Use, Mobility and Environment* 1, 91–104.
Corbisiero, F. and Paura, R. (2020) *Turismo*. Naples: Italian Institute for the Future.
Corbisiero, F. and Monaco, S. (2021) Post-pandemic tourism resilience: Changes in Italians' travel behavior and the possible responses of tourist cities. *Worldwide Hospitality and Tourism Themes* 13 (3), 401–417. doi: 10.1108/WHATT-01-2021-0011.
Corbisiero, F. and Monaco, S. (2022, in press) Homosexual tourism: An ideal model of 'sustainable rainbow tourist destination'. In M. Novelli, C. Milano and J.M. Cheer (eds) *Handbook of Niche Tourism*. Cheltenham: Edward Elgar.
Cordeniz, J.A. (2002) Recruitment, retention, and management of generation X: A focus on nursing professionals. *Journal of Healthcare Management* 47 (4), 237–249.
Correia, A. and Dolnicar, S. (eds) (2021) *Women's Voices in Tourism Research – Contributions to Knowledge and Letters to Future Generations*. Brisbane: The University of Queensland. See https://uq.pressbooks.pub/tourismknowledge (accessed 20 October 2021).
Costa, C., Gilliland, G. and McWatt, J. (2019) 'I want to keep up with the younger generation' – older adults and the web: A generational divide or generational collide? *International Journal of Lifelong Education* 38 (5), 566–578.
Costa, P. and Lopes, R. (2013) Urban design, public space and creative milieus: An international comparative approach to informal dynamics in cultural districts. *Cidades, Comunidades e Territórios* 26, 40–66.
Cotter, D., Hermsen, J.M. and Vanneman, R. (2011) The end of the gender revolution? Gender role attitudes from 1977 to 2008. *American Journal of Sociology* 117 (1), 259–289.
Couldry, N. and McCarthy, A. (2003) *MediaSpace: Place, Scale and Culture in a Media Age*. London: Routledge.
Coupland, D. (1991) *Generation X: Tales for an Accelerated Culture*. New York: St. Martin's Press.
Crespi, I. and Ruspini, E. (eds) (2016) *Balancing Work and Family in a Changing Society: The Fathers' Perspective*. Basingstoke: Palgrave MacMillan.
Cross, G. (2018) *Machines of Youth: America's Car Obsession*. Chicago, IL: University of Chicago Press.
Crouch, D., Jackson, R. and Thompson, F. (eds) (2005) *The Media and the Tourist Imagination: Converging Cultures*. London: Routledge.
CSIS-IYF (2017) 2016 global millennial viewpoints survey. See https://iyfglobal.org/sites/default/files/library/2016-Global-Millenial-Viewpoints-Survey.pdf (accessed 30 March 2020).
Cushman and Wakefield (2020) Demographic shift: The world in 2030. See https://www.cushmanwakefield.com/en/insights/demographic-shifts-the-world-in-2030 (accessed 20 October 2020).

Das, S.S. and Tiwari, D.A. (2020) Understanding international and domestic travel intention of Indian travellers during Covid-19 using a Bayesian approach. *Tourism Recreation Research* 1, 1–17.

Davis, J.B., Pawlowski, S.D. and Houston, A. (2006) Work commitments of baby boomers and Gen-Xers in the IT profession: Generational differences or myth? *Journal of Computer Information Systems* 46 (3), 43–49.

De Seta, C. (1993) *L'Italia del Grand Tour: da Montaigne a Goethe*. Naples: Electra.

Deal, J.J., Stawiski, S., Graves, L.M., Gentry, W.A., Ruderman, M. and Weber, T.J. (2012) Perceptions of authority and leadership: A cross-national, cross-generational investigation. In E.S. Ng, S.T. Lyons and L. Schweitzer (eds) *Managing the New Workforce: International Perspectives on the Millennial Generation* (pp. 281–306). Cheltenham: Edward Elgar.

Deem, R. (1982) Women, leisure and inequality. *Leisure Studies* 1 (1), 29–46.

Deep Focus (2018) *Pinterest Consumers Study*. New York: Deep Focus' Intelligence Group.

Dell Technologies (2018) Gen Z: The future has arrived. See https://www.delltechnologies.com/content/dam/digitalassets/active/en/unauth/sales-documents/solutions/gen-z-the-future-has-arrived-executive-summary.pdf (accessed 15 July 2020).

Deloitte (2015) Collaboration generation. See https://www2.deloitte.com/content/dam/Deloitte/lu/Documents/technology/lu_en_collaboration-generation_122015.pdf (accessed 15 July 2020).

Deloitte (2017) The 2017 Deloitte millennial survey. See https://www2.deloitte.com/content/dam/Deloitte/ru/Documents/about-deloitte/en/millennials-report-global-2017-en.pdf (accessed 15 July 2020).

Deloitte (2018) The 2018 Deloitte millennial survey. See https://www2.deloitte.com/content/dam/Deloitte/global/Documents/About-Deloitte/gx-2018-millennial-survey-report.pdf (accessed 15 July 2020).

Deloitte (2019) The 2019 Deloitte global millennial survey. See https://www2.deloitte.com/global/en/pages/about-deloitte/articles/millennialsurvey.html (accessed 15 July 2020).

Deloitte (2020) The 2020 Deloitte global millennial survey. See https://www2.deloitte.com/global/en/pages/about-deloitte/articles/millennialsurvey.html (accessed 15 July 2020).

Deloitte (2021) The 2021 Deloitte global millennial and Gen Z survey. See https://www2.deloitte.com/content/dam/Deloitte/global/Documents/2021-deloitte-global-millennial-survey-report.pdf (accessed 15 September 2021).

Demartini, J.R. (1985) Change agents and generational relationships: A re-evaluation of Mannheim's Problem of Generations. *Social Forces* 64 (1), 1–16.

Dencker, J.C., Joshi, A. and Martocchio, J.J. (2008) Towards a theoretical framework linking generational memories to workplace attitudes and behaviours. *Human Resource Management Review* 18 (3), 180–187.

Derber, C. (2015) *Sociopathic Society: A People's Sociology of the United States*. London: Routledge.

Dermott, E. (2008) *Intimate Fatherhood: A Sociological Analysis*. London/New York: Routledge.

Desai, P.R., Desai, P.N., Ajmera, K.D. and Mehta, K. (2014) A review paper on Oculus Rift: A virtual reality headset. *IJETT Journal* 13 (4), 175–179.

Diepstraten, I., Ester, P. and Vinken, H. (1999) Talkin' 'bout my generation: Ego and alter images of generations in the Netherlands. *The Netherlands' Journal of Social Sciences* 35 (2), 91–109.

Dilthey, W. (1875) *Gesammelte Schriften*, V, 37. Berlin: Verlag der Königl. Akademie der Wissenschaften.

Dilthey, W. (1910) *Der Aufbau der geschichtlichen Welt in den Geisteswissenschaften Erstdruck* [*The Formation of the Historical World in the Human Sciences*]. Berlin: Verlag der Königl. Akademie der Wissenschaften.

Dimitriou, C.K. and AbouElgheit, E. (2019) Understanding generation Z's social decision-making in travel. *Tourism and Hospitality Management* 25 (2), 311–334.

Di Nicola, P. and Ruspini, E. (2020) Family and family relations at the time of Covid-19: An introduction. *Italian Sociological Review* 10 (3S), 679–685.

Dolan, B. (2001) *Ladies of the Grand Tour: British Women in Pursuit of Enlightenment and Adventure in Eighteenth-Century Europe*. New York: HarperCollins Publishers.

Donati, P. (1995) Ripensare le generazioni e il loro intreccio. *Studi di Sociologia* 33 (3), 203–223.

Donzelli, G., Cioni, L., Cancellieri, M.G., Llopis Morales, A. and Morales Suárez-Varela, M.M. (2020) The effect of the Covid-19 lockdown on air quality in three Italian medium-sized cities. *Atmosphere* 11, 1118.

Doucet, A. (2006) *Do Men Mother? Fathering, Care and Domestic Responsibility*. Toronto: University of Toronto Press.

Drescher, J. (2015) Out of DSM: Depathologizing homosexuality. *Behavioural Sciences* 5 (4), 565–575.

Duffett, R.G. (2015) Facebook advertising's influence on intention-to-purchase and purchase amongst millennials. *Internet Research* 25 (4), 498–526.

Duffy, B., Shrimpton, H. and Clemence, M. (2017) *Millennial Myths and Realities*. London: IPSOS Mori.

Dunn, M. (2019) *Millennials for America*. New York: Lulu.com.

Durant, I. and Coke-Hamilton, P. (2020) Covid-19 requires gender-equal responses to save economies. UNCTAD 1 April. See https://unctad.org/en/pages/newsdetails.aspx?OriginalVersionID=2319 (accessed 6 April 2020).

Dutton, S. (2018) *The Post-Experience Economy: Travel in an Age of Sameness*. Düsseldorf: Euromonitor International.

The Economist (2017) Sustainable investment joins the mainstream Millennials are coming into money and want to invest it responsibly. See: https://www.economist.com/finance-and-economics/2017/11/25/sustainable-investment-joins-the-mainstream (accessed 2 February 2022).

Edmunds, J. and Turner, B.S. (2002) *Generations, Culture and Society*. Maidenhead: Open University Press.

Edmunds, J. and Turner, B.S. (2005) Global generations: Social change in the twentieth century. *The British Journal of Sociology* 56 (4), 559–577.

EEA (2021) *Urban Sustainability in Europe. Opportunities in Challenging Times*. Vienna: European Economic Association.

Eisenstadt, S.N. (1956) *From Generation to Generation*. New York: The Free Press of Glencoe.

Eisenstadt, S.N. (1963) *The Political Systems of Empires*. New York: The Free Press of Glencoe.

Eisner, S.P. (2005) Managing Generation Y. *SAM Advanced Management Journal* 70, 4–15.

Elliott, C. and Reynolds, W. (2019) *Making it Millennial*. New York: Deloitte University Press.

Elwalda, A., Lü, K. and Ali, M. (2016) Perceived derived attributes of online customer reviews. *Computer Human Behaviour* 56, 306–319.

ETC – European Travel Commission (2020) *Study on Generation Z Travellers*. Brussels: ETC Market Intelligence. See https://etc-corporate.org/reports/study-on-generation-z-travellers/ (accessed 10 November 2020).

Euromonitor International (2019) *Top 10 Global Consumer Trends*. Düsseldorf: Euromonitor International.

Euromonitor International (2020) *The Rise of Vegan and Vegetarian Food*. Düsseldorf: Euromonitor International.

European Commission (2017) *Reflection Paper on Harnessing Globalization*. Brussels: European Commission. See https://ec.europa.eu/commission/sites/beta-political/files/reflection-paper-globalisation_en.pdf (accessed 18 September 2020).

European Commission (2020) *The Digital Economy and Society Index (DESI)*. Brussels: European Commission.

European Parliament (2015) *Research for Tran Committee. Tourism and the Sharing Economy: Challenges and Opportunities for the UE*. Brussels: European Parliament, Policy Department for Structural and Cohesion Policies.

European Travel Commission (2020) Generation Z recognises its responsibility in shaping the future of travel in Europe. See https://etc-corporate.org/news/generation-z-recognises-its-responsibility-in-shaping-the-future-of-travel-in-europe/ (accessed 18 November 2022).

Eurosif (2019) *Eurosif 2018 SRI Study*. Brussels: Eurosif.

Eurostat (2020) Tourism in the EU: What a normal spring season looks like - before Covid-19. See https://ec.europa.eu/eurostat/statistics-explained/index.php?title=Tourism_in_the_EU_-_what_a_normal_spring_season_looks_like_-_before_Covid-19&stable=1 (accessed 1 March 2021).

Expedia (2018) *A Look Ahead: How Younger Generations are Shaping the Future of Travel*. Seattle, WA: Expedia Group Media Solution.

Expedia Group Media Solutions (2019) Gen Alpha and family travel trends. See https://www.vatc.org/wp-content/uploads/2020/01/Generation-Alpha-and-Family-Trends-2019-by-Expedia-Group.pdf (accessed 18 September 2020).

Eyerman, R. and Turner, B.S. (1998) Outline of a theory of generations. *European Journal of Social Theory* 1 (1), 91–106.

Fagiani, M.L. (2010) Turismo LGBT. In E. Marra and E. Ruspini (eds) *Altri turismi: Viaggi, esperienze, emozioni* (pp. 85–100). Milan: FrancoAngeli.

Felson, M. and Spaeth, J.L. (1978) Community structure and collaborative consumption: A routine activity approach. *American Behavioural Scientist* 21 (4), 614–624.

Femenia-Serra, F., Neuhofer, B. and Baidal, J. (2019) Towards a conceptualisation of smart tourists and their role within the smart destination scenario. *Service Industries Journal* 39, 109–133.

Fiani, C.N. and Han, H.J. (2019) Navigating identity: Experiences of binary and non-binary transgender and gender non-conforming (TGNC) adults. *International Journal of Transgenderism* 20 (2–3), 181–194.

Figueroa-Domecq, C. and Segovia-Pérez, M. (2020) Application of a gender perspective in tourism research: A theoretical and practical approach. *Journal of Tourism Analysis: Revista de Análisis Turístico* 27 (2), 251–270.

Figueroa-Domecq, C., Pritchard, A., Segovia-Pérez, M., Morgan, N. and Villace-Molinero, T. (2015) Tourism gender research: A critical accounting. *Annals of Tourism Research* 52, 87–103.

Finger, M., Bert, N., Kupfer, D., Montero, J.J. and Wolek, M. (2017) *Research for TRAN Committee: Infrastructure Funding Challenges in the Sharing Economy*. Brussels: European Parliament, Policy Department for Structural and Cohesion Policies.

Fondevila-Gascón, J.F., Berbel, G. and Muñoz-González, M. (2019) Experimental study on the utility and future of collaborative consumption platforms offering tourism related services. *Future Internet* 11 (80).

France, A. and Roberts, S. (2015) The problem of social generations: A critique of the new emerging orthodoxy in youth studies. *Journal of Youth Studies* 18 (2), 215–230.

Frändberg, L. and Vilhelmson, B. (2011) More or less travel: Personal mobility trends in the Swedish population focusing gender and cohort. *Journal of Transport Geography* 19, 1235–1244.

Friedemann, M-L. and Buckwalter, K. (2014) Family caregiver role and burden related to gender and family relationships. *Journal of Family Nursing* 20 (3), 313–336.

Friedman, A.T. (1985) The influence of humanism on the education of girls and boys in Tudor England. *History of Education Quarterly* 25 (1/2), 57–70. doi: 10.2307/368891.
Friskopp, A. and Silverstein, S. (1996) *Straight Jobs Gay Lives*. New York: Simon and Schuster.
Furr, H.L., Bonn, M.A. and Hausman, A. (2001) A generational and geographical analysis of internet travel-services usage. *Tourism Analysis* 6 (2), 139–147.
Furstenberg, F. (2017) The use and abuse of millennials as an analytic category. See https://contemporaryfamilies.org/8-furstenberg-millennials-analytic-category (accessed 18 March 2020).
Fuse (2020) How Gen Z is reacting to Covid-19 and 22 ways brands can take action right now. See https://www.fusemarketing.com/thought-leadership/genz-reaction-covid-19-ways-brands-can-act-now (accessed 16 April 2020).
Galland, D. (2009) Interview with N. Howe. *Casey Research* 36–47.
Gardiner, S., King, C. and Grace, D. (2013) Travel decision making: An empirical examination of generational values, attitudes, and intentions. *Journal of Travel Research* 52 (3), 310–324.
Gardiner, S., Grace, D. and King, C. (2014) The generation effect: The future of domestic tourism in Australia. *Journal of Travel Research* 53 (6), 705–720.
Gardiner, S., Grace, D. and King, C. (2015) Is the Australian domestic holiday a thing of the past? Understanding baby boomer, Generation X and Generation Y perceptions and attitude to domestic and international holidays. *Journal of Vacation Marketing* 21 (4), 336–350.
Garikapati, V., Pendyala, R., Morris, E., Mokhtarian, P. and McDonald, N. (2016) Activity patterns, time use, and travel of millennials: A generation in transition? *Transport Reviews* 36, 558–584.
GBD (2017) *Global Burden of Diseases*. Washington, DC: University of Washington.
GenForward (2018) *Millennials' View on LGBT Issues*. New York: GenForward Survey.
Gharzai, L.A., Beeler, W.H. and Jagsi, R. (2020) Playing into stereotypes: Engaging millennials and Generation Z in the Covid-19 pandemic response. *Advances in Radiation Oncology* 5 (4), 679–681.
Giddens, A. (1990) *The Consequences of Modernity*. Cambridge: Polity Press.
Giddens, A. (1993) *Sociology*. Cambridge: Polity Press.
Ginsberg, C. (2017) The market for vegetarian foods. *Age* 8, 18.
Gilleard, C. (2004) Cohorts and generations in the study of social change. *Social Theory and Health* 2 (1), 106–119.
Gillespie, N. (2014) Millennials are selfish and entitled, and helicopter parents are to blame. *Time* 21 August.
Gilli, M. (2009) *Autenticità e interpretazione nell'esperienza turistica*. Milan: FrancoAngeli.
Gilli, M. (2015) *Turismo e identità*. Naples: Liguori.
Gilli, M. and Ruspini, E. (2014) What is old and what is new? Representations of masculinity in travel brochures. In T. Thurnell-Read and M. Casey (eds) *Men, Masculinities, Travel and Tourism* (pp. 204–218). Basingstoke: Palgrave MacMillan.
Gillis, J.R. (1974) *Youth and History: Tradition and Change in European Age Relations, 1770–Present*. New York: Academic Press.
Gilovich, T. and Kumar, A. (2015) We'll always have Paris: The hedonic payoff from experiential and material investments. *Advances in Experimental Social Psychology* 51, 147–187.
Gilovich, T., Kumar, A. and Jampol, L. (2014) A wonderful life: Experiential consumption and the pursuit of happiness. *Journal of Consumer Psychology* 25 (1), 152–165.
Giorgetti Fumel, M. (2010) *Legami virtuali. Internet: dipendenza o soluzione?* Trapani: Di Girolamo.
Gleadhill, E. (2017) Performing travel: Lady Holland's grand tour souvenirs and the house of all Europe. In J. Milam (ed.) *Cosmopolitan Moments: Instances of Exchange in the*

Long Eighteenth Century, emaj – Electronic Melbourne Art Journal (Special Issue) 9.1, December.

Global Web Index (2019) *2019 Social Media User Trends Report*. London: Global Web Index.

Global Web Index (2020) *Coronavirus: The Big Consumer Shift so Far*. London: Global Web Index.

GlobalData (2020) *Global Data Survey*. London: GlobalData UK Ltd.

Glover, P. (2009) Generation Y's future tourism demand: Some opportunities and challenges. In P. Benckendorff, G. Moscardo and D. Pendergast (eds) *Tourism and Generation Y* (pp. 155–164). Wallingford: CABI International.

Glover, P. and Prideaux, B. (2006) The impact of demographic change on future tourism demand: A focus group study. CAUTHE 2006 Conference. See https://www.researchgate.net/publication/43497749_The_impact_of_demographic_change_on_future_tourism_demand_a_focus_group_study (accessed 16 August 2020).

Godtman-Kling, K., Margaryan, L. and Fuchs, M. (2020) (In)equality in the outdoors: Gender perspective on recreation and tourism media in the Swedish mountains. *Current Issues in Tourism* 23 (2), 233–247.

Gössling, S., Scott, D. and Hall, C.M. (2020) Pandemics, tourism and global change: A rapid assessment of Covid-19. *Journal of Sustainable Tourism* 29 (5).

Gouldner, A.W. (1960) *The Norm of Reciprocity: A Preliminary Statement*. Washington, DC: Washington University.

Green, E., Hebron, S. and Woodward, D. (1990) A social history of women's leisure. In E. Green, S. Hebron and D. Woodward (eds) *Women's Leisure, What Leisure?* (pp. 38–56). London: Palgrave MacMillan.

Gretzel, U. (2006) Consumer generated content: Trends and implications for branding. *e-Review of Tourism Research* 4 (3), 9–11.

Gretzel, U., Sigala, M., Xiang, Z. and Koo, C. (2015) Smart tourism: Foundations and developments. *Electronic Markets* 25 (3), 179–188.

Guaracino, J. and Salvato, E. (2017) *Handbook of LGBT Tourism and Hospitality: A Guide for Business Practice*. New York: Columbia University Press.

Gyr, U. (2010) *The History of Tourism: Structures on the Path to Modernity Tourism. European History Online* (EGO). Mainz: Institute of European History (IEG).

Haddouche, H. and Salomone, C. (2018) Generation Z and the tourist experience: Tourist stories and use of social networks. *The Journal of Tourism Futures – ETFI* (Special Issue) 4 (1), 69–79.

Hall, D., Swain, M. and Kinnaird, V. (2003) Tourism and gender: An evolving agenda. *Tourism Recreation Research* 28 (2), 7–11.

Hamari, J., Sjöklint, M. and Ukkonen, A. (2015) The sharing economy: Why people participate in collaborative consumption. *Journal of the Association for Information Science and Technology* 1, 1–28.

Han, H., Xu, H. and Chen, H. (2018) Social commerce: A systematic review and data synthesis. *Electronic Commerce Research Applied* 30, 38–50.

Hansen, J-I.C. and Leuty, M.E. (2012) Work values across generations. *Journal of Career Assessment* 20 (1), 34–52.

Hanson, S. (2010) Gender and mobility: New approaches for informing sustainability. *Gender, Place and Culture* 17 (1), 5–23.

Hantrais, L. (2004) *Family Policy Matters: Responding to Family Change in Europe*. Bristol: The Policy Press.

Hargreaves, J. (1989) The promise and problems of women's leisure and sport. In C. Rojek (ed.) *Leisure for Leisure* (pp. 130–149). London: Palgrave MacMillan.

Harmony, M. (2020) 'Coronnials'. *Urban Dictionary*. See https://www.urbandictionary.com/define.php?term=coronnials (accessed 18 November 2022).

Hartal, G. (2020) Touring and obscuring: How sensual, embodied and haptic gay touristic practices construct the geopolitics of pinkwashing. *Social and Cultural Geography* 1, 1–19.
Haydam, N., Purcărea, T., Edu, T. and Negricea, C. (2017) Explaining satisfaction at a foreign tourism destination: An intra-generational approach evidence within Generation Y from South Africa and Romania. *Amfiteatru Economic* 19 (45), 528–542.
Haynie, A. (2014) Self-transformation through dangerous travel: Mary Morris's nothing to declare and Audrey Schulman's the cage. In G.R. Ricci (ed.) *Travel, Discovery, Transformation* (pp. 20–32). New Brunswick, NJ/London: Transaction Publishers.
Hayward, P. (2002) *GCSE Leisure and Tourism for OCR*. Oxford: Einemann.
Heelas, L. and Woodhead, P. (2005) *The Spiritual Revolution: Why Religion Is Giving Way To Spirituality*. Hoboken, NJ: Blackwell Publishing.
Henderson, K.A. (1991) The contribution of feminism to an understanding of leisure constraints. *Journal of Leisure Research* 23 (4), 363–377.
Hendl, T., Chung, R. and Wild, V. (2020) Pandemic surveillance and racialized subpopulations: Mitigating vulnerabilities in Covid-19 apps. *Journal of Bioethical Inquiry* 17 (4), 829–834.
Herbig, P., Koehler, W. and Day, K. (1993) Marketing to the baby bust generation. *Journal of Consumer Marketing* 10 (1), 4–9.
Hersant, Y. (1988) *Italies. Anthologies des voyageurs français au XVIII et XIX siècles*. Paris: Robert Laffont.
Hicks, R. and Hicks, K. (1999) *Boomers, Xers and Other Strangers: Understanding the General Differences that Divide Us*. Wheaton, IL: Tyndale House.
Holcomb, B. and Luongo, M. (1996) Gay tourism in the United States. *Annals of Tourism Research* 23 (3), 695–726.
Horak, S. and Weber, S. (2000) Youth tourism in Europe: Problems and prospects. *Tourism Recreation Research* 25, 37–44.
Horwath HTL (2015) Tourism megatrends. 10 things you need to know about the future of tourism. See http://corporate.cms-horwathhtl.com/wp-content/uploads/sites/2/2015/12/Tourism-Mega-Trends4.pdf (accessed 3 December 2020).
Hostels.com (2017a) *Mobile Travel Tracker*. Dublin: Hostelworld Group PLC.
Hostels.com (2017b) *Tourist Attraction for Millennials*. Dublin: Hostelworld Group PLC.
Hout, M. (2019) Social mobility. *Pathways*, State of The Union. Millennial Dilemma (Special Issue), 29–32.
Howe, N. and Strauss, W. (1991) *Generations: The History of America's Future, 1584 to 2069*. New York: HarperCollins.
Howe, N. and Strauss, W. (2000) *Millennials Rising: The Next Great Generation*. New York: Vintage Books.
Howe, N. and Strauss, W. (2008) *Millennials Go to College: Strategies for a New Generation on Campus*. Great Falls, VA: LifeCourse Associates.
Huang, W. and Lu, Y. (2017) Generational perspective on consumer behaviour: China's potential outbound tourist market. *Tourism Management Perspectives* 24, 7–15.
Huang, Y.-C. and Petrick, J. F (2010) Generation Y's travel behaviours: A comparison with baby boomers and generation X. In P. Benckendorff, G. Moscardo and D. Pendergast (eds) *Tourism and Generation Y* (pp. 27–37). Cambridge: CABI International.
Hughes, H.L. (1997) Holidays and homosexual identity. *Tourism Management* 18 (1), 3–7.
Hughes, H.L. (2002) Gay men's holiday destination choice: A case of risk and avoidance. *International Journal of Tourism Research* 4 (4), 299–312.
Hughes, H.L. (2006) *Pink Tourism: Holidays of Gay Men and Lesbians*. Oxford: CABI.
Hughes, H.L. and Deutsch, R. (2010) Holidays of older gay men: Age or sexual orientation as decisive factors? *Tourism Management* 31 (4), 454–463.

Hughes, H.L., Monterrubio, C. and Miller, A. (2010) Gay tourists and host community attitudes. *International Journal of Tourism Research* 12 (6), 774–786.

Hui-Chun, Y. and Miller, P. (2003) The generation gap and cultural influence: A Taiwan empirical investigation. *Cross-Cultural Management* 10, 23–41.

Hui-Chun, Y. and Miller, P. (2005) Leadership style: The X Generation and baby boomers compared in different cultural contexts. *Leadership and Organisation Development* 26, 35–50.

Hunnicutt, B. and Pine, B.J. (2020) *The Age of Experiences: Harnessing Happiness to Build a New Economy*. Philadelphia, PA: Temple University Press.

Hussain, A. and Fusté-Forné, F. (2021) Post-pandemic recovery: A case of domestic tourism in Akaroa (South Island, New Zealand). *World* 2 (1), 127–138.

Ianole-Calin, R., Druica, E., Hubona, G. and Wu, B. (2020) What drives Generations Y and Z towards collaborative consumption adoption? Evidence from a post-communist environment. *Kybernetes* 2020.

IEEE (2019) Generation AI 2019 survey. How millennial parents are embracing health and wellness Technologies for their generation alpha kids. See https://transmitter.ieee.org/health-2019/?_ga=2.153959800.1930124194.1604768984-482560373.1604768984 (accessed 10 September 2020).

IGLTA (2020) *IGLTA Annual Guide 2020*. Fort Lauderdale, FL: International LGBTQ+ Travel Association.

ILGTA (2019) *State-Sponsored Homophobia Report*. Geneva: The International Lesbian, Gay, Bisexual, Trans and Intersex Association.

Inglehart, R. (1977) *The Silent Revolution. Changing Values and Political Styles among Western Publics*. Princeton, NJ: Princeton University Press.

Insead, Head Foundation and Universum (2014) Understanding a misunderstood generation. The first large-scale survey of how millennials attitudes and actions vary across the globe, and the implication for employers. See https://headfoundation.org/wp-content/uploads/2020/11/thf-papers_Understanding-a-misunderstood-generation (accessed 18 March 2020).

Ioannides, Y.M. and Rosenthal, S.S. (1995) Estimating the consumption and investment demands for housing and their effect on housing tenure status. *Financial Services Review* 4 (1), 62–75.

IOP (2019) *Harvard IOP Youth Poll 2019*. Harvard, MA: Harvard University Press.

Ipsos Mori (2018) Gen Z. Beyond binary. See https://thinks.ipsos-mori.com/category/gen-z-beyond-binary/ (accessed 16 April 2020).

iResearch (2020) *The Penetration Rate of China's Used Car E-Commerce*. Shanghai: iResearch.

Ivanov, S. (2017) Robonomics: Principles, benefits, challenges, solutions. *Yearbook of Varna University of Management* 10, 283–293.

Ivars-Baidal, J.A., Celdrán-Bernabeu, M.A., Mazón, J.N. and Perles-Ivars, Á.F. (2017) Smart destinations and the evolution of ICTs: A new scenario for destination management?. *Current Issues in Tourism* 22 (13), 1581–1600.

Jackson, E.L. and Henderson, K.A. (1995) Gender-based analysis of leisure constraints. *Leisure Sciences* 17 (1), 31–51.

Jackson, T. (2009) *Prosperity without Growth: Economics for a Finite Planet*. London: Earthscan.

Jacobsen, J.P., Khamis, M. and Yuksel, M. (2015) Convergences in men's and women's life patterns: Lifetime work, lifetime earnings, and human capital investment. *Research in Labor Economics* 41, 1–33.

Jacobsen, M.H. and Tester, K. (2012) *Utopia: Social Theory and the Future*. Burlington, VT: Ashgate.

Jaeger, J. (1985) Generations in history: Reflections on a controversial concept. *History and Theory* 24 (3), 273–292 (originally published 1977 in *Geschichte und Gesellschaft*).

Jansen, N. (1974) Definition of generation and sociological theory. *Social Science* 49 (2), 90–98.

Jaskulsky, L. and Besel, R. (2013) Words that (don't) matter: An exploratory study of four climate change names in environmental discourse. *Applied Environmental Education and Communication* 12 (1), 38–45.

Javalgi, R.G., Thomas, E.G. and Rao, S.R. (1992) Consumer behaviour in the US pleasure travel marketplace: An analysis of senior and nonsenior travellers. *Journal of Travel Research* 31 (1), 14–19.

Jenkins, F. and Smith, J. (2021) Work-from-home during Covid-19: Accounting for the care economy to build back better. *The Economic and Labour Relations Review* 32 (1), 22–38.

Jesus, T.S., Landry, M.D. and Jacobs, K. (2020) A 'new normal' following COVID-19 and the economic crisis: Using systems thinking to identify challenges and opportunities in disability, telework, and rehabilitation. *Work* 67 (1), 37–46.

Jordan, E. (1991) 'Making good wives and mothers'? The transformation of middle-class girls' education in nineteenth-century Britain. *History of Education Quarterly* 31 (4), 439–462.

Jordan, F. and Gibson, H. (2005) 'We're not stupid... but we'll not stay home either': Experiences of solo women travellers. *Tourism Review International* 9 (2), 195–212.

Joshi, A., Dencker, J., Franz, G. and Martocchio, J. (2010) Unpacking generational identities in organizations. *Academy of Management Review* 35 (3), 392–414.

Judt, T. (2005) *Postwar. A History of Europe since 1945*. New York: The Penguin Press.

JWT-J. Walter Thompson Intelligence in partnership with Snap Inc (2019) Into Z future understanding Generation Z, the next generation of super creatives. See https://assets.ctfassets.net/inb32lme5009/5DFlqKVGIdmAu7X6btfGQt/44fdca09d7b630ee28f5951d54feed71/Into_Z_Future_Understanding_Gen_Z_The_Next_Generation_of_Super_Creatives_.pdf (accessed 28 October 2020).

Kardes, F., Cronley, M. and Cline, T. (2014) *Consumer Behaviour*. Mason, OH: Cengage Learning.

Kelly, P.F. (2012) Labor, movement: Migration, mobilities and geographies of work. In T.J. Barnes, J. Peck and E. Sheppard (eds) *The Wiley-Blackwell Companion to Economic Geography* (pp. 431–443). Hoboken, NJ: Wiley Blackwell.

Kenway, J. and Epstein, D. (2021) The Covid-19 conjuncture: Rearticulating the school/home/work nexus. *International Studies in Sociology of Education* 1.

Kertzer, D.I. (1983) Generation as a sociological problem. *Annual Review of Sociology* 9, 125–149.

Khan, S. (2011) Gendered leisure: Are women more constrained in travel for leisure? *Tourismos* 6 (1), 105–121.

Khoo-Lattimore, C. and Wilson, E. (2017) *Women and Travel: Historical and Contemporary Perspectives*. Waretown, NJ: Apple Academic Press.

Khoshpakyants, A. and Vidischcheva, E. (2012) *Challenges of Youth Tourism*. Sochi: State University for Tourism and Recreation.

Kinnaird, V. and Hall, D. (2000) Theorizing gender in tourism research. *Tourism Recreation Research* 25 (1), 71–84. doi: 10.1080/02508281.2000.11014901.

Kinnaird, V., Kothari, U. and Hall, D. (1994) Tourism: Gender perspectives. In V. Kinnaird and D. Hall (eds) *Tourism: A Gender Analysis* (pp. 1–34). Chichester: John Wiley and Sons.

Knight, Y. (2009) Talkin' 'bout my generation: A brief introduction to generational theory. *Planet* 21 (1), 13–15.

Knopp, L. (1992) Sexuality and the spatial dynamics of capitalism. *Environment and Planning D: Society and Space* 10 (6), 651–669.

Koczanski, P. and Rosen, H.S. (2019) *Are Millennials Really So Selfish? Preliminary Evidence from the Philanthropy Panel Study*. Princeton, NJ: Griswold Center for Economic Policy Studies.

Konrath, S., O'Brien, E. & Hsing. (2011). Changes in Dispositional Empathy in American College Students Over Time: A Meta-Analysis. *Personality and social psychology review: an official journal of the Society for Personality and Social Psychology* 15, 180–98. doi: 10.1177/1088868310377395.

Kortti, J. (2011) Generations and media history. In L. Fortunati and F. Colombo (eds) *Broadband Society and Generational Changes, Series: Participation in Broadband Society*, Vol. 5 (pp. 69–93). Frankfurt am Main: Peter Lang.

Kroløkke, C. and Sørenson, A.S. (2005) Three waves of feminism: From suffragettes to Grrls. In C. Kroløkke and A.S. Sørenson (eds) *Gender Communication Theories and Analyses. From Silence to Performance* (pp. 1–23). Thousand Oaks, CA: Sage.

Kupperschmidt, B.R. (2000) Multigenerational employees: Strategies for effective management. *Health Care Manager* 19, 65–76.

Kurz, C., Li, G. and Vine, D.J. (2018) Are millennials different? *Finance and Economics Discussion Series* 2018 (080). doi: 10.17016/FEDS.2018.080

Lab42 (2019) *What's Mine is Yours... and Yours... and Yours*. Chicago, IL: Lab42 Research.

Lamanna, M.A. and Riedmann, A. (2009) *Marriages and Families: Making Choices in a Diverse Society*. Belmont, CA: Thomson Wadsworth.

Lamb, S. (2015) Generation in anthropology. In J.D. Wright (ed.) *International Encyclopedia of the Social and Behavioural Sciences* (Vol. 9; 2nd edn; pp. 853–856). Oxford: Elsevier.

Lancaster, L.C. and Stillman, D. (2002) *When Generations Collide: Who They Are. Why They Clash. How to Solve the Generational Puzzle at Work*. New York: HarperCollins.

Lang, M., Aguinaga, M., Mokrani, D. and Santillana, A. (2013) Development critiques and alternatives: A feminist perspective. In M. Lang and D. Mokrani (eds) *Beyond Development. Alternative Visions from Latin America Publisher* (pp. 41–60). Quito-Ecuador: Transnational Institute, Rosa Luxemburg Foundation.

Leach, R., Phillipson, C., Biggs, S. and Money, A. (2013) Baby boomers, consumption and social change: The bridging generation? *International Review of Sociology* 23 (1), 104–122.

Lee, D. and Lee, J. (2020) Testing on the move: South Korea's rapid response to the Covid-19 pandemic. *Transportation Research Interdisciplinary Perspectives* 5, 100–111.

Lehto, X., Jang, S., Achana, F. and O'Leary, J. (2008) Exploring tourism experience sought: A cohort comparison of baby boomers and the silent generation. *Journal of Vacation Marketing* 14 (3), 237–252.

Leung, D., Law, R., Van Hoof, H. and Buhalis, D. (2013) Social media in tourism and hospitality: A literature review. *Journal of Travel & Tourism Marketing* 30 (1–2), 3–22.

Lewis, L.A. (ed.) (1992) *The Adoring Audience: Fan Culture and Popular Media*. New York/London: Routledge.

Li, X. and Wang, Y. (2011) China in the eyes of western travellers as represented in travel blogs. *Journal of Travel and Tourism Marketing* 28 (7), 689–719.

Li, X., Li, X.R. and Hudson, R. (2013) The application of generational theory to tourism consumer behaviour: An American perspective. *Tourism Management* 37, 147–164.

Li, Y. (2019) *Li Yinhe Talks About Love*. Beijing: Beijing October Literature and Arts Publishing House.

Lin, J.H., Lin, J-H., Lee, S-J., Yeh, C., Lee, W-H. and Wong, J-Y. (2014) Identifying gender differences in destination decision making. *Journal of Tourism and Recreation* 1, 1–11.

Lindeman, C.K. (2017) *Representing Duchess Anna Amalia's Bildung: A Visual Metamorphosis*. Abingdon/New York: Routledge.

Little, C., Patterson, D., Moyle, B. and Bec, A. (2018) Every footprint tells a story: 3D scanning of heritage artifacts as an interactive experience. *Association for Computing Machinery Proceedings of the Australasian Computer Science Week Multiconference* 38, 1–8.

Littlewood, I. (2001) *Sultry Climates: Travel and Sex Since the Grand Tour*. London: John Murray.

Liu, Y. (2019) *Millennials' Attitudes Towards Influencer Marketing And Purchase Intentions*. Los Angeles, CA: California State University.

Lorber, J. (1994) *Paradoxes of Gender*. New Haven, CT: Yale University Press.

Lord, T. (2021) *Covid-19 and Climate Change: How to Apply the Lessons of the Pandemic to the Climate Emergency*. London: Tony Blair Institute for Global Change.

Losyk, B. (1997) How to manage an X'er. *The Futurist* 31, 43.

Lu, J.L. (2009) Effect of work intensification and work extensification on women's health in the globalised labour market. *Journal of International Women's Studies* 10 (4), 111–126.

Lyons, M., Lavelle, K. and Smith, D. (2017) *Gen Z Rising*. New York: Accenture Strategy.

Lyons, S. and Kuron, L. (2014) Generational differences in the workplace: A review of the evidence and directions for future research. *Journal of Organizational Behaviour* 35, 139–157.

Mackay, H. (1997) *Generations, Baby Boomers, Their Parents and Their Children*. Sydney: MacMillan.

Magvadkar, A., White, O., Krishnan, M., Mahajan, D. and Azcue, X. (2020) *Covid-19 and Gender Equality: Countering the Regressive Effects*. New York: McKinsey Global Institute.

Mandich, G. (1996) *Spazio-tempo. Prospettive sociologiche*. Milan: FrancoAngeli.

Mannheim, K. (1928) Das problem der generationen. *Kölner Vierteljahres Hefte für Soziologie* 157–184.

Mannheim, K. (1952) The problem of generations. In P. Kecskemeti (ed.) *Essays on the Sociology of Knowledge* (pp. 276–320). London: Routledge and Kegan Paul.

Marías, J. (1970) *Generations, a Historical Method*. Alabama, MS: The University of Alabama Press.

Marra, E. and Ruspini, E. (eds) (2010) *Altri turismi. Viaggi, esperienze, emozioni*. Milan: FrancoAngeli.

Martin, J.C. and Lewchuk, W. (2018) The Generation Effect. Millennials, Employment Precarity and the 21st Century Workplace. McMaster University and (PEPSO) – Project on Poverty and Employment Precarity in Southern Ontario. See https://www.economics.mcmaster.ca/pepso/documents/the-generation-effect-full-report.pdf (accessed 18 November 2021).

Martínez-López, F. and D'Alessandro, S. (eds) (2020) *Advances in Digital Marketing and eCommerce*. New York: Springer Proceedings in Business and Economics.

Matthews-Sawyer, M., McCullough, K. and Myers, P. (2002) Maiden voyages: The rise of women-only travel. *PATA Compass Magazine* July–August, 36–40.

McCrindle, M. and Wolfinger, E. (2009) *The ABC of XYZ: Understanding the Global Generations*. Sydney: New South Wales.

McDonald, N.C. (2015) Are millennials really the 'GoNowhere' generation? *Journal of the American Planning Association* 81 (2), 90–103.

McEwan, C. (2000) *Gender, Geography and Empire: Victorian Women Travellers in West Africa*. Aldershot: Ashgate Publishing.

McGinnis, J., Chun, S. and McQuillan, J. (2003) A review of gendered consumption in sport and leisure. *Academy of Marketing Science Review* 5, 1–24.

McKinsey (2018) *True Gen: Generation Z and its Implications for Companies*. New York: McKensey and Company.

McNamara, K.E and Prideaux, B. (2010) A typology of solo independent women travellers. *International Journal of Tourism Research* 12, 253–264.

Mcquaid, R.W. and Chen, T. (2012) Commuting times: The role of gender, children and part time work. *Research in Transportation Economics* 34, 66–73.

Melián-González, A., Moreno-Gil, S. and Araña, J.E. (2011) Gay tourism in a sun and beach destination. *Tourism Management* 32 (5), 1027–1037.

Meltz, R. (2017) Growing up with Alexa. What will it do to kids to have digital butlers they can boss around? *MIT Technology Review* 8.

Mentré, F. (1920) *Les générations sociales*. Paris: éd. Bossard.

Meredith, G. and Schewe, C. (1994) The power of cohorts. *American Demographics* 16 (12), 22–31.

Merico, M. (2012) Giovani, generazioni e mutamento nella sociologia di Karl Mannheim. *Studi di Sociologia* 50 (1), 109–129.

Migacz, S.N.D and Petrick, J. (2018) Millennials: America's cash cow is not necessarily a herd. *Journal of Tourism Futures* 4 (1), 16–30.

Milano, C. and Koens, K. (2021) The paradox of tourism extremes: Excesses and restraints in times of Covid-19. *Current Issues in Tourism* 1, 1–13.

Miller, D.S. (2021) Abrupt new realities amid the disaster landscape as one crisis gives way to crises. *Worldwide Hospitality and Tourism Themes* 13 (3), 304–311.

Millett, K. (1971) *Sexual Politics*. London: Granada Publishing.

Mitchell, R. (2002) The Generation Game: Generation X and Baby Boomer Wine Tourism. Paper presented at a meeting of the New Zealand Tourism and Hospitality Research, Rotorua, New Zealand.

Monaco, S. (2018a) Mobilità turistiche fuori dai luoghi. Forme e significati dei viaggi online per i giovani italiani. *Fuori Luogo. Journal of Sociology of Territory, Tourism, Technology* 4 (2), 91–104.

Monaco, S. (2018b) Tourism and the new generations: Emerging trends and social implications in Italy. *The Journal of Tourism Futures – ETFI* (Special Issue) 4 (1), 7–15.

Monaco, S. (2019a) Mixed methods e e-research: frontiere possibili per lo studio delle hidden population. *Sociologia Italiana – AIS Journal of Sociology* 14, 97–108.

Monaco, S. (2019b) *Sociologia del turismo accessibile. Il diritto alla mobilità e alla libertà di viaggio*. Velletri: PM Editore.

Monaco, S. (2021) *Tourism, Safety and Covid-19 Security, Digitization and Tourist Behavior*. New York: Routledge.

Monaco, S. and Nothdurfter, U. (2021) Stuck under the rainbow? Gay parents' experiences with transnational surrogacy and family formation in times of Covid-19. *Italian Sociological Review* 11 (2), 509–530.

Monterrubio, C. (2018) Tourism and male homosexual identities: Directions for sociocultural research. *Tourism Review* 74 (5), 1058–1069.

Monterrubio, C., Rodriguez-Madera, S. and Pérez Díaz, J. (2020) Trans women in tourism: Motivations, constraints and experiences. *Journal of Hospitality and Tourism Management* 43 (6), 169–178.

Moores, S. (2012) *Media, Place and Mobility*. New York: Palgrave MacMillan.

Morace, F. (2016) *ConsumAutori. I nuovi nuclei generazionali*. Milan: Egea.

Moreira, M. and Campos, L. (2019) The ritual of ideological interpellation in LGBT tourism and the impossibility of the desire that moves. *Brazilian Magazine of Pesquisa em Turismo* 13 (2), 54–68.

Morena, E., Krause, D. and Stevis, D. (2020) *Just Transitions: Social Justice in a Low-Carbon World*. London: Pluto.

Moreno, A. and Urraco, M. (2018) The generational dimension in transitions: A theoretical review. *Societies MDPI* 8 (3), 1–12.

Morgan, B. (2019) NOwnership, no problem: An updated look at why millennials value experiences over owning things. *Forbes* 1.

Morris, M. and Western, B. (1999) Inequality in earnings at the close of the twentieth century. *Annual Review of Sociology* 25, 623–657. See http://www.jstor.org/stable/223519.

MSCI (2020) Millennials. Demographic change and the impact of a generation, thematic insights. See https://www.msci.com/documents/1296102/17292317/ThematicIndex-Millennials-cbr-en.pdf/44668168-67fd-88cd-c5f7-855993dce7c4?t=1587390986253 (accessed 20 October 2020).

Muchnick, M. (1996) *Naked Management: Bare Essentials for Motivating the X-Generation at Work*. Delray Beach, FL: St Lucie.

Munar, A.M. (2017) To be a feminist in (tourism) academia. *Anatolia* 28 (4), 514–529.

Murphy, E.F., Gordon, J.D. and Anderson, T.L. (2004) Cross-cultural, cross-cultural age and cross-cultural generational differences in values between the United States and Japan. *Journal of Applied Management and Entrepreneurship* 9, 21–47.

Murray, D.A.B. (2007) The civilized homosexual: Travel talk and the project of gay identity. *Sexualities* 10 (1), 49–60.

Nast, H.J. (2002) Queer patriarchies, queer racism, international. *Antipode* 35 (5), 874–909.

Neuhofer, B., Buhalis, D. and Ladkin, A. (2012) Conceptualising technology enhanced destination experiences. *Journal of Destination Marketing & Management* 1, 36–46.

Nielsen (2017) Young and ready to travel (and shop). A look at millennial travellers. See https://v-i-r.de/wp-content/uploads/2017/02/nielsen-millennial-traveller-study-jan-2017-1.pdf (accessed 18 August 2020).

Nielsen (2019) *Nielsen Global Generational Lifestyles Survey*. New York: Nielsen.

Nigam, R., Pandya, K., Luis, A.J., Sengupta, R. and Kotha, M. (2020) Positive effects of Covid-19 lockdown on air quality of industrial cities (Ankleshwar and Vapi) of Western India. *Scientific Reports* 11, 4285.

Nocifora, E. (2008) *La società turistica*. Naples: Scriptaweb.

Novelli, M. (2005) *Niche Tourism: Contemporary Issues, Trends and Cases*. Oxford: Butterworth-Heinemann Ltd.

Nyìri, K. (2005) *The Local and the Global in Mobile Communication*. Vienna: Passagen Verlag.

OAS–CIM Organization of American States–The Inter-American Commission of Women (2020) *Covid-19 in Women's Lives: Reasons to Recognize the Differential Impacts*. General Secretariat of the Organization of the American States (GS/OAS).

Obenour, W., Patterson, M. and Pedersen, P. (2004) Conceptualization of a meaning-based research approach for tourism service experiences. *Tourism Management* 27 (1), 34–41.

OC&C (2019) *A Generation Without Borders: Embracing Generation Z*. New York: OC&C Strategy Consultants.

Ofcom (2016) *Children and Parents: Media Use and Attitudes Report 2016*. London: Ofcom.

Oláh, L.S., Kotowska, I.E. and Richter, R. (2018) The new roles of men and women and implications for families and societies. In G. Doblhammer and J. Gumà (eds) *A Demographic Perspective on Gender, Family and Health in Europe* (pp. 41–64). Cham: Springer.

Olcelli, L. (2015) Lady Anna Riggs Miller: The 'modest' self-exposure of the female grand tourist. *Studies in Travel Writing* 19 (4), 312–323.

Olson, E. and Reddy Best, K.L. (2019) Transgender and gender non-conforming individuals and the negotiation of identity development through embodied practices while traveling: Panopticism and gendered surveillance. *International Textile and Apparel Association Annual Conference Proceedings* 76 (1).

ONSM (2019) *Rapporto Nazionale sulla Sharing Mobility*. Rome: Osservatorio nazionale sulla sharing mobility.
Oppenheim Mason, J. and Jensen, A.-M. (eds) (1995) *Gender and Family Change in Industrialized Countries*. New York: Clarendon Press.
Oppermann, M. (1995) Travel life cycle. *Annals of Tourism Research* 22 (3), 535–552.
Ortega y Gassett, J. (1933) *The Modern Theme*. London: Harper and Row.
Osservatorio e-commerce B2C (2020) *Netcomm Forum Report*. Milan: Politecnico di Milano.
Oswick, C., Grant, D. and Oswick, R. (2020) Categories, crossroads, control, connectedness, continuity, and change: A metaphorical exploration of Covid-19. *The Journal of Applied Behavioural Science* 56 (3), 284–288.
Ottman, J. (2011) *The New Rules of Green Marketing: Strategies, Tools, and Inspiration for Sustainable Branding*. San Francisco, CA: Taylor & Francis Ltd.
Özkan, M. and Solmaz, B. (2015) The changing face of the employees: Generation Z and their perceptions of work (a study applied to university students). *Procedia Economics and Finance* 26, 476–483.
Pai, M. (2020) Covidization of research: What are the risks? *Nature Medicine* 26 (8), 1159.
Pan, B., MacLaurin, T. and Crotts, J.C. (2007) Travel blogs and their implications for destination marketing. *Journal of Travel Research* 46 (1), 35–45.
Parker, K. and Livingston, G. (2018) 8 Facts about American dads. Pew Research Center, Washington, DC. 13 June. See https://www.pewresearch.org/fact-tank/2018/06/13/fathers-day-facts/ (accessed 10 September 2020).
Parker, K. and Igielnik, R. (2020) On the cusp of adulthood and facing an uncertain future: What we know about Gen Z so far. Pew Research Center Social and Demographic Trends. 14 May. See https://www.pewsocialtrends.org/essay/on-the-cusp-of-adulthood-and-facing-an-uncertain-future-what-we-know-about-gen-z-so-far/ (accessed 10 August 2020).
Parker, K., Graf, N. and Igielnik, R. (2019) Generation Z looks a lot like millennials on key social and political issues. Pew Research Center Social and Demographic Trends. 17 January. See https://www.pewsocialtrends.org/2019/01/17/generation-z-looks-a-lot-like-millennials-on-key-social-and-political-issues/ (accessed 9 April 2020).
Parry, E. and Urwin, P. (2011) Generational differences in work values: A review of theory and evidence. *International Journal of Management Reviews* 13, 79–96.
Patterson, K., Grenny, J. and Maxfield, D. (2013) *Influencer: The New Science of Leading Change*. New York: McGraw-Hill Education.
Pecorelli, V. (2016) Padri Digitali: come gestire la gravidanza e la paternità online (Digital Fathers). In E. Ruspini, M. Inghilleri and V. Pecorelli (eds) *Diventare padri nel Terzo Millennio [Fathers in the Third Millennium]* (pp. 97–109). Milan: FrancoAngeli.
Pederson, E.B. (1992) Future seniors: Is the hospitality industry ready for them? *FIU Hospitality Review* 10 (2), 1–8.
Pemble, J. (1987) *The Mediterranean Passion: Victorians and Edwardians in the South*. Oxford: Clarendon Press.
Pendergast, D. (2010) Getting to know the Y generation. In P. Benckendorff, G. Moscardo and D. Pendergast (eds) *Tourism and Generation Y* (pp. 85–97). Wallingford: CAB International.
Pennington-Gray, L.A. and Kerstetter, D.L. (2001) What do university-educated women want from their pleasure travel experiences? *Journal of Travel Research* 40 (1), 49–56.
Pennington-Gray, L. and Lane, C. (2001) Profiling the silent generation. *Journal of Hospitality and Leisure Marketing* 9 (1–2), 73–95.
Pennington-Gray, L., Kerstetter, D.L. and Warnick, R. (2002) Forecasting travel patterns using Palmore's cohort analysis. *Journal of Travel and Tourism Marketing* 13 (1–2), 125–143.

Pennington-Gray, L., Fridgen, J.D. and Stynes, D. (2003) Cohort segmentation: An application to tourism. *Leisure Sciences* 25 (4), 341–361.

Pereira, A. and Silva, C. (2018) Women solo travellers: Motivations and experiences. *Millennium – Journal of Education, Technologies, and Health* 2 (6), 99–106.

Perkins, K.M., Munguia, N., Ellenbecker, M., Moure-Eraso, R. and Velazquez, L. (2021) Covid-19 pandemic lessons to facilitate future engagement in the global climate crisis. *Journal of Cleaner Production* 290, 125178.

Perra, M.S. and Ruspini, E. (eds) (2013) Men who work in 'non-traditional' occupations. *International Review of Sociology – Revue Internationale de Sociologie*, Themed Section/Section Thématique 23 (2). See https://www.tandfonline.com/toc/cirs20/23/2

Perzanowski, A. and Schultz, J. (2016) *The End of Ownership: Personal Property in the Digital Economy*. Cambridge: MIT Press.

Pew Research Center (2015) The whys and hows of generations research. 3 September. See https://www.pewresearch.org/politics/2015/09/03/the-whys-and-hows-of-generations-research/ (accessed 3 November 2020).

PhoCusWright (2020) *Europe Online 2020*. New York: Northstar Travel Media LLC.

Pilia, A. (2019) *Rapporto Aniasa 2019: numeri e trend del noleggio auto*. Rome: Associazione Nazionale Industria dell'Autonoleggio.

Pilcher, J. (1994) Mannheim's sociology of generations: An undervalued legacy. *The British Journal of Sociology* 45 (3), 481–495. https://doi.org/10.2307/591659.

Pinder, W. (1926) *Das Problem der Generation in der Kunstgeschichte Europas*. Munich: Bruckmann.

Pine, B.J. and Gilmore, J.H. (2001) *The Experience Economy*. Cambridge, MA: Harvard Business School Press.

Pine, B.J. and Gilmore, J.H. (2019) *The Experience Economy: Competing for Customer Time, Attention, and Money*. Cambridge, MA: Harvard Business School Press.

Piotrowski, J.T. and Valkenburg, P. (2017) *Plugged In: How Media Attract and Affect Youth*. New Haven, CT: Yale University Press.

Plummer, K. (1981) *The Making of the Modern Homosexual*. Totowa, NJ: Barnes and Noble.

Pooley, C.G., Turnbull, J. and Adams, M. (2005) *A Mobile Century? Changes in Everyday Mobility in Britain in the Twentieth Century*. Aldershot: Ashgate.

Poria, Y. (2006) Assessing gay men and lesbian women's hotel experiences: An exploratory study of sexual orientation in the travel industry. *Journal of Travel Research* 44 (3), 327–334.

Priporas, C.V., Stylos, N. and Fotiadis, A.K. (2017) Generation Z consumers' expectations of interactions in smart retailing: A future agenda. *Computers in Human. Behaviour* 77, 374–381.

Pritchard, A. (2018) Predicting the next decade of tourism gender research. *Tourism Management Perspectives* 25, 144–146.

Pritchard, A. and Morgan, N.J. (2000) Privileging the male gaze: Gendered tourism landscapes. *Annals of Tourism Research* 27 (4), 884–905.

Pritchard, A. and Morgan, N.J. (2017) Tourism's lost leaders: Analysing gender and performance. *Annals of Tourism Research* 63, 34–47.

Pritchard, A., Morgan, N.J., Sedgley, D., Khan, E. and Jenkins, A. (2000) Sexuality and holiday choices: Conversations with gay and lesbian tourists. *Leisure Studies* 19 (4), 267–282.

Pritchard A., Morgan, N. and Sedgley, D. (2002) In search of lesbian space? The experience of Manchester's gay village. *Leisure Studies* 21 (2), 105–123.

Pritchard, A., Morgan, N.J., Ateljevic, I. and Harris, C. (eds) (2007) *Tourism and Gender: Embodiment, Sensuality and Experience*. Wallingford: CABI Publishing.

Puar, J.K. (2002) Circuits of queer mobility: Tourism, travel and globalization. *GLQ, A Journal of Lesbian and Gay Studies* 8 (1–2), 101–137.

Purhonen, S. (2016) Generations on paper: Bourdieu and the critique of 'generationalism'. *Social Science Information* 55 (1), 94–114.

PwC (2011) Millennials at work. Reshaping the workplace. See https://www.pwc.de/de/prozessoptimierung/assets/millennials-at-work-2011.pdf (accessed 18 March 2020).

PwC (2015) The female millennial: A new era of talent. See https://www.pwc.com/jg/en/publications/the-female-millennial_a-new-era-of-talent.pdf (accessed 18 March 2020).

Rainer, T. and Rainer, J. (2011) *The Millennials: Connecting to America's Largest Generation*. Nashville, TN: BandH Publishing Group.

Ram, Y., Kama, A., Mizrachi, I. and Hall, C.M. (2019) The benefits of an LGBT-inclusive tourist destination. *Journal of Destination Marketing and Management* 14, 10–16.

Ranzini, G., Newlands, G., Anselmi, G., Andreotti, A., Eichhorn, T., Etter, M., Hoffmann, C., Jürss, S. and Lutz, C. (2015) *Millennials and the Sharing Economy: European Perspectives*. Amsterdam: Free University of Amsterdam.

Resonance Consultancy (2018) *World's Best Cities. A Ranking of Global Place Equity*. New York: Resonance Consultancy.

Richards, G. (ed.) (2007) *Cultural Tourism: Global and Local Perspectives*. New York: Haworth.

Richards, G. and Morrill, W. (2020). Motivations of global millennial travellers. *Revista Brasileira de Pesquisa em Turismo* 14 (1), 126–139.

Richter, L.K. (1994) Exploring the political role of gender in tourism research. In W.F. Theobald (ed.) *Global Tourism: The Next Decade* (pp. 391–404). Oxford: Butterworth Heinemann.

Ridgeway, C.L. and Smith-Lovin, L. (1999) The gender system and interaction. *Annual Review of Sociology* 25, 191–216.

Rifkin, J. (2011) *The Third Industrial Revolution: How Lateral Power is Transforming Energy, the Economy, and the World*. New York: Palgrave MacMillan.

Risman, B.J. (2004) Gender as a social structure: Theory wrestling with activism. *Gender and Society* 18 (4), 429–450.

Risman, B.J. (2018) *Where the Millennials Will Take Us: A New Generation Wrestles with the Gender Structure*. New York: Oxford University Press.

Robertson, R. (1992) *Globalization: Social Theory and Global Culture*. London: Sage.

Robinson, J. (1990) *Wayward Women: A Guide to Women Travellers*. London: Oxford University Press.

Robinson, J. (1994) *Unsuitable for Ladies: An Anthology of Women Travellers*. Oxford: Oxford University Press.

Robinson, V.M. and Schänzel, H.A. (2019) A tourism inflex: Generation Z travel experiences. *Journal of Tourism Futures* 5 (2), 127–141.

Rodríguez-Urrego, D. and Rodríguez-Urrego, L. (2020) Air quality during the Covid-19: PM2.5. Analysis in the 50 most polluted capital cities in the world. *Environmental Pollution* 266, 115042.

Rogers, R. and Botsman, R. (2010) *What's Mine is Yours: The Rise of Collaborative Consumption*. New York: HarperCollins.

Rojek, C. and Urry, J. (eds) (1997) *Touring Cultures: Transformations of Travel and Theory*. London: Routledge

Rokka, J. (2010) Netnographic inquiry and new translocal sites of the social. *International Journal of Consumer Studies* 34 (4), 381–387.

Rosenbloom, S. (2006) *Understanding Women's and Men's Travel Patterns Research on Women's Issues in Transportation: The Research Challenge, in Transportation Research Board, Research on Women's Issues in Transportation*. Report of a Conference, Volume 1, The National Academies Press, Washington, DC, 7–28. See https://www.nap.edu/read/23274/chapter/4 (accessed 11 November 2021).

Rubin, G. (1975) The traffic in women: Notes on the 'political economy' of sex. In R.R. Reyter (ed.) *Toward an Anthropology of Women* (pp. 157–210). New York: Monthly Review Press.

Ruspini, E. (2019) Millennial men, gender equality and care: The dawn of a revolution? *Teorija in Praksa* (Special Issue) 56 (4), 985–1000.

Ruspini, E., Hearn, J., Pease, B. and Pringle, K. (eds) (2011) *Men and Masculinities Around the World. Transforming Men's Practices*. Basingstoke: Palgrave MacMillan.

Russell, G. and Bohan, J. (2005) The gay generation gap: Communicating across the divide. *Angles* 8 (1), 1–8.

Ruting, B. (2008) Economic transformations of gay urban spaces: Revisiting Collins' evolutionary gay district model. *Australian Geographer* 39 (3), 259–269.

Ryan, B. (1992) *Feminism and the Women's Movement: Dynamics of Change in Social Movement Ideology and Activism*. New York: Routledge.

Rybcyzynski, W. (1991) *Waiting for the Weekend*. London: Viking Penguin.

Sajjadi, A., Åkesson Castillo, L.C.F. and Sun, B. (2012) Generational differences in work attitudes: A comparative analysis of Generation Y and preceding generations from companies in Sweden. Thesis. Internationella Handelshögskolan, Högskolan i Jönköping, Sweden.

Sambuco, P. (ed.) (2015) *Italian Women Writers, 1800–2000: Boundaries, Borders, and Transgression*. Madison, WI: Fairleigh Dickinson University Press.

Schewe, C.D. and Noble S.M. (2000) Market segmentation by cohorts: The value and validity of cohorts in America and abroad. *Journal of Marketing Management* 16 (1–3), 129–142.

Schewe, C.D. and Meredith, G. (2004) Segmenting global markets by generational cohorts: Determining motivations by age. *Journal of Consumer Behaviour* 4 (1), 51–63.

Schiffman, L., Bednall, D., O'Cass, A., Paladino, A., Ward, S. and Kanuk, L. (2008) *Consumer Behaviour*. Frenchs Forest: Pearson Education Australia.

Schofields Insurance (2017) *Holiday Destination Chosen Based on How 'Instagrammable' the Holiday Pics Will Be*. Bolton: Schofields Ltd.

Schor, J.B. and Fitzmaurice, C.J. (2015) Collaborating and connecting: The emergence of the sharing economy. In L. Reisch and J. Thogersen (eds) *Handbook on Research on Sustainable Consumption* (pp. 410–425). Cheltenham: Edward Elgar.

Schulz, K. (2008) The women's movement. In M. Klimke and J. Scharloth (eds) *1968 in Europe* (pp. 281–293). Palgrave MacMillan: New York.

Schuman, H. and Scott, J. (1989) Generations and collective memories. *American Sociological Review* 54, 359–381.

Scott, J. (1986) Gender: A useful category of historical analysis. *American Historical Review* 91, 1053–1075.

Seemiller, C. and Grace, M. (2016) *Generation Z Goes to College*. San Francisco, CA: Jossey-Bass.

Segovia-Pérez, M., Figueroa-Domecq, C., Moraleda, L. and Muñoz, A. (2019) Incorporating a gender approach in the hospitality industry: Female executives' perceptions. *International Journal of Hospitality Management* 76 (A), 184–193.

Sengupta, D. (2017) *The Life of Y: Engaging Millennials as Employees and Consumers*. New Delhi: Sage Publications India.

Sezgin, E. and Yolal, M. (2012) Golden age of mass Tourism: Its history and development, visions for global tourism industry – Creating and Sustaining Competitive Strategies, Murat Kasimoglu, IntechOpen. See https://www.intechopen.com/books/visions-for-global-tourism-industry-creating-and-sustaining-competitive-strategies/mass-tourism-its-history-and-development-in-the-golden-age (accessed 31 March 2020).

Sfodera, F. (2012) *Turismi, Destinazioni e Internet. La rilevazione della consumer experience nei portali turistici*. Milan: FrancoAngeli.

Shaheen, S.A., Chan, N.D., Bansal, A. and Cohen, A. (2015) Shared mobility: Definitions, industry developments, and early understanding. *Innovative Mobility Research* 11, 1–27.

Shaw, S.M. (1994) Gender, leisure and constraint: Toward a framework for analysis of women's leisure. *Journal of Leisure Research* 26 (1), 8–22.

Shaw, S.M. (2001) Conceptualizing resistance: Women's leisure as political practice. *Journal of Leisure Research* 33 (2), 186–201.
Shores, K., Scott, D. and Floyd, M.F. (2007) Constraints to outdoor recreation: A multiple hierarchy stratification perspective. *Leisure Sciences* 29 (3), 227–246.
Shridhar, A. (2019) *Millennial Parents Transforming Family Life.* London: Euromonitor International.
Siegel, L. (2011) *Homo interneticus. Restare umani nell'era dell'ossessione digitale.* Prato: Piano B.
Simirenko, A. (1966) Mannheim's generational analysis and acculturation source. *The British Journal of Sociology* 17 (3), 292–299.
Simmel, G. (1950) *The Sociology of Georg Simmel.* Glencoe: Free Press.
Singer, P. and Prideaux, B. (2006) The impact of demographic change on future tourism demand: A focus group study. In P.A. Whitelaw and G.B. O'Mahony (eds) *CAUTHE 2006: To the City and Beyond* (pp. 336–345). Footscray, Vic.: Victoria University, School of Hospitality, Tourism and Marketing.
Sivak, M. and Schoettle, B. (2016) *Recent Decreases in the Proportion of Persons with a Driver's License across All Age Groups.* Ann Arbor, MI: University of Michigan Transportation Research Institute.
Skinner, H., Sarpong, D. and White, G.R. (2018) Meeting the needs of the millennials and Generation Z: Gamification in tourism through geocaching. *Journal of Tourism Futures* 4 (1), 93–104.
Smith, A. (2016) *Shared, Collaborative and On Demand: The New Digital Economy.* Washington, DC: Pew Research Center. Internet and Technology.
Smith, A.G. (2020) *Marketing to the New Generations of LGBTQ+ Tourists.* Hattiesburg, MS: University of Southern Mississippi.
Sofronov, B. (2018) Millennials: A new trend for the tourism industry. *Annals of Spiru Haret University* 18, 109–122.
Sorokin, P.A. (1947) *Society, Culture, and Personality.* New York: Harper.
Sosa, K. (2019) The Damron address book, a green book for gays, kept a generation of men in the know. *Los Angeles Magazine* 6, 25.
Southan, J. (2017) From Boomers to Gen Z: Travel trends across the generations. *Globetrender Magazine,* 19 May. See https://globetrender.com/2017/05/19/travel-trends-across-generations/ (accessed 8 August 2020).
Sparks and Honey (2019) *Generation Z 2025: The Final Generation.* New York: Sparks and Honey.
Spitzer, A.B. (1973) The historical problem of generations. *The American Historical Review* 78 (5), 1353–1385.
Staffieri, S. (2016) *L'esperienza turistica dei giovani italiani.* Rome: Sapienza Università Editrice
Statnickė, G. (2019) An expression of different generations in an organization: A systematic literature review. *Society. Integration. Education, Proceedings of the International Scientific Conference,* 5, 273–291.
Strauss, W. and Howe, N. (1991) The cycle of generations. *American Demographics* 13 (4), 24–33.
Strauss, W. and Howe, N. (1992) *Generations: The History of America's Future, 1584 to 2069.* New York: Harper.
Strauss, W. and Howe, N. (1997) *The Fourth Turning. An American Prophecy. What Cycles of History Tell Us about America's Next Rendezvous with Destiny.* New York: Broadway Books.
Stuart Mill, J. (1865) *A System of Logic.* London: Longmans.
Stuber, M. (2002) Tourism marketing aimed at gay men and lesbians: A business perspective. In S. Clift, M. Luongo and C. Callister (eds) *Gay Tourism: Culture, Identity and Sex* (pp. 88–124). New York: Continuum.
Swain, M. (1995) Gender in tourism. *Annals of Tourism Research* 22 (2), 247–266.

Swain, M. (1995) Gender in tourism. *Annals of Tourism Research* 22 (2), 247–266.
Swain, M. and Henshall Momsen, J. (eds) (2002) *Gender/Tourism/Fun*. Elmsford, NY: Cognizant Communication Corp.
Swan, M. (2012) Sensor mania! The internet of things, wearable computing, objective metrics, and the quantified self 2.0. *Journal of Sensor and Actuator Networks* 1, 217–253.
Swan, M. (2013) The quantified self: Fundamental disruption in big data science and biological discovery. *Big Data* 1, 85–99.
SWG (2019) *Lotta contro i cambiamenti climatici*. Trieste: SWG Spa.
Syngellakis, S., Probstl-Haider, U. and Pineda, F. (2018) *Sustainable Tourism*. Southampton: WIT.
Szarycz, G.S. (2008) Cruising, freighter-style: A phenomenological exploration of tourist recollections of a passenger freighter travel experience. *International Journal of Tourism Research* 10 (3), 259–269.
Tannenbaum, E.R. (1976) *1900: The Generation before the Great War*. Garden City, NY: Anchor Press.
Tapscott, D. (2008) *Grown Up Digital: How the Net Generation is Changing Your World*. New York: McGraw-Hill.
Tavares, J., Sawant, M. and Ban, O. (2018) A study of the travel preferences of generation Z located in Belo Horizonte (Minas Gerais – Brazil). *e-Review of Tourism Research (eRTR)* 15 (2–3), 223–241.
Taylor, P. and Keeter, S. (eds) (2010) Millennials: A portrait of generation next. Confident, connected, open to change. Pew Research Center, Washington DC. February. See http://www.pewresearch.org/wp-content/uploads/sites/3/2010/10/millennials-confident-connected-open-to-change.pdf (accessed 23 March 2020).
Telefónica (2013) *Global Millennial Survey. Global Results*. See https://milunesco.unaoc.org/mil-resources/telefonica-global-millennial-survey-global-results/ (accessed 18 March 2020).
The Council of Economic Advisers (2014) *15 Economic Facts about Millennials*. See https://obamawhitehouse.archives.gov/sites/default/files/docs/millennials_report.pdf (accessed 18 March 2020).
Therkelsen, A., Blichfeldt, B.S., Chor, J. and Ballegaard, N. (2013) 'I am very straight in my gay life': Approaching an understanding of lesbian tourists' identity construction. *Journal of Vacation Marketing* 19 (4), 317–327.
Thevenot, G. (2007) Blogging as a social media. *Tourism and Hospitality Research* 7 (3–4), 287–289.
Thiefoldt, D. and Scheef, D. (2004) Generation X and the millennials: What you need to know about mentoring the new generations. Law Practice Today. See http://www.abanet.org/lpm/lpt/articles/nosearch/mgt08044_print.html (accessed 23 March 2020).
Thijs, P., Te Grotenhuis, M., Scheepers, P. and van den Brink, M. (2019) The rise in support for gender egalitarianism in the Netherlands, 1979–2006: The roles of educational expansion, secularization, and female labor force participation. *Sex Roles* 81 (9–10), 594–609.
Thompson, M.E. and Armato, M. (2012) *Investigating Gender*. Cambridge: Polity Press.
Tilley, S. and Houston, D. (2016) The gender turnaround: Young women now travelling more than young men. *Journal of Transport Geography* 54, 349–358.
Topdeck Travel (2020) *Topdeck Travel Survey 2020*. London: Topdeck Travel.
Towner, J. (1994) Tourism history: Past, present and future. In A.V. Seaton (ed.) *Tourism: The State of the Art* (pp. 721–728). Chichester: John Wiley and Sons.
Transportation Research Board (2006) *Research on Women's Issues in Transportation*. Report of a Conference, Volume 1, The National Academies Press, Washington, DC. See https://www.nap.edu/read/23274/chapter/4 (accessed 18 November 2021).
Tribe, J. (2006) The truth about tourism. *Annals of Tourism Research* 33 (2), 360–381.

Trinidad, J.E. (2021) Equity, engagement, and health: School organisational issues and priorities during Covid-19. *Journal of Educational Administration and History* 53 (1), 67–80.

TripAdvisor (2015) *TripBarometer Connected Traveler*. New York: TripAdvisor, Inc.

Trua, T. (2016) *Sharing economy. Economia della condivisione*. Bologna: Bitbiblos.

Tussyadiah, I. (2020) A review of research into automation in tourism. *Annals of Tourism Research* 81.

Tussyadiah, I. and Fesenmaier, D.R. (2009) Mediating tourist experiences: Access to places via shared videos. *Annals of Tourism Research* 36, 24–40.

Tussyadiah, I. and Inversini, A. (2015) *Information and Communication Technologies in Tourism*. London: Springer.

Tuttle, B. (2014) Can we stop pretending the sharing economy is all about sharing? *Money* 1, 14–30.

Twenge, J.M. (2013) The evidence for Generation me and against Generation we. *Emerging Adulthood* 1 (1), 11–16.

Tzuo, T. and Weisert, G. (2018) *Subscribed: Why the Subscription Model Will Be Your Company's Future and What to Do About It*. New York: The Penguin Press.

UN Women (2020) Covid-19: Emerging gender data and why it matters. See https://data.unwomen.org/resources/covid-19-emerging-gender-data-and-why-it-matters (accessed 6 April 2020).

UNFPA – United Nations Population Fund (2020) Coronavirus Disease (Covid-19) Gender Equality and Addressing Gender-based Violence (GBV) and Coronavirus Disease (Covid-19) Prevention, Protection and Response. UNFPA Interim Technical Brief. See https://www.unfpa.org/resources/gender-equality-and-addressing-gender-based-violence-gbv-and-coronavirus-disease-covid-19 (accessed 6 April 2020).

UN – United Nations (2020a) Covid-19 and Transforming Tourism. Policy Brief August. See https://reliefweb.int/sites/reliefweb.int/files/resources/policy-brief-the-impact-of-covid-19-on-women-en.pdf (accessed 15 September 2020).

UN – United Nations (2020b) The Impact of Covid-19 on Women. Policy Brief April. See https://reliefweb.int/sites/reliefweb.int/files/resources/policy-brief-the-impact-of-covid-19-on-women-en.pdf (accessed 28 April 2020).

UNWTO – UN World Tourism Organization (2011a) *Global Report on Women in Tourism 2010*. Madrid: WTO. See https://www.e-unwto.org/doi/pdf/10.18111/9789284413737 (accessed 6 April 2020).

UNWTO – UN World Tourism Organization (2011b) *Tourism Highlights 2011 edition*. Madrid: WTO.

UNWTO – UN World Tourism Organization (2017) *Second Global Report on LGBT Tourism*. Madrid: WTO.

UNWTO – UN World Tourism Organization (2019) *Global Report on Women in Tourism, Second edition*. Madrid: WTO. See https://www.e-unwto.org/doi/book/10.18111/9789284420384 (accessed 6 April 2020).

UNWTO – UN World Tourism Organization (2020a) Covid-19. Putting people first. See https://www.unwto.org/tourism-covid-19 (accessed 20 April 2020).

UNWTO – UN World Tourism Organization (2020b) Impact assessment of the Covid-19 outbreak on international tourism. See https://www.unwto.org/impact-assessment-of-the-covid-19-outbreak-on-international-tourism (accessed 15 December 2020).

UNWTO – UN World Tourism Organization (2020c) *World Tourism Barometer* 18 (5), August–September. See https://www.e-unwto.org/doi/epdf/10.18111/wtobarometereng.2020.18.1.5 (accessed 15 October 2020).

UNWTO – UN World Tourism Organization (2020d) Covid-19 and vulnerable groups. See https://www.unwto.org/covid-19-inclusive-response-vulnerable-groups (accessed 5 August 2020).

Urry, J. (2003) Social networks, travel and talk. *British Journal of Sociology* 54 (2), 155–175.

Urry, J. and Sheller, M. (2004) *Tourism Mobilities: Places to Play, Places in Play*. London: Routledge.

Urry, J. and Larsen, J. (2011) *The Tourist Gaze 3.0*. London: Sage.

Valentine, G. (1989) The geography of women's fear. *Area* 21, 385–390.

Valentine, G. and Skelton, T. (2003) Finding oneself, losing oneself: The lesbian and gay scene as a paradoxical space. *International Journal of Urban and Regional Research* 27 (4), 849–866.

van Dijk, J. (1991) *De Netwerk maatschappij, Sociale Aspecten van Nieuwe Media*. Alphen aan den Rijn: Samsom.

van Gils, W. and Kraaykamp, G. (2008) The emergence of dual-earner couples: A longitudinal study of the Netherlands. *International Sociology* 23 (3), 345–366.

Varkey Foundation (2019) *Generation Z: Global Citizenship Survey*. New York: Varkey Foundation.

Veiga, C., Custódio Santos, M., Águas, P. and Santos, J.A.C. (2017) Are millennials transforming global tourism? Challenges for destinations and companies. *Worldwide Hospitality and Tourism Themes* 9 (6), 603–616.

Verboven, H. and Vanherck, L. (2016) The sustainability paradox of the sharing economy. *UWF* 24, 303–314.

Veríssimo, M. and Costa, C. (2018) Do hostels play a role in pleasing millennial travellers? The Portuguese case. *Journal of Tourism Futures* 4 (1), 57–78.

Vincent, J.A. (2005) Understanding generations: Political economy and culture in an ageing society. *The British Journal of Sociology* 56 (4), 579–599.

Virtuoso (2015) Best of the best award nominees. See https://www.virtuoso.com/getmedia/9ec307b0-acb8-4059-bb49-0e058ba52d6a/Best-of-the-Best-Nominees-FINAL.aspx (accessed 18 November 2021).

Vizcaino-Suárez, P., Jeffrey, H. and Eger, C. (eds) (2020) *Tourism and Gender-based Violence. Challenging Inequalities*. Wallingford: CABI Publishing.

Vogt, P.W. (1997) *Tolerance and Education: Learning to Live with Diversity and Difference*. Thousand Oaks, CA: Sage Publications.

Vorobjovas-Pinta, O. (ed.) (2021) *Gay Tourism: New Perspectives*. Bristol: Channel View Publications.

Vukonić, B. (2012) An outline of the history of tourism theory. In C.H.C. Hsu and W.C. Gartner (eds) *The Routledge Handbook of Tourism Research* (pp. 3–27). Abingdon/New York: Routledge.

Wagar, W.W. (1983) H.G. Wells and the genesis of future studies. *WNRF* 1, 25–29.

Waitt, G. (2003) Gay games: Performing 'community' out from the closet of the locker room. *Society and Cultural Geography* 4 (2), 167–183.

Waitt, G. and Markwell, K. (2006) *Gay Tourism: Culture and Context*. London: Routledge.

Walrave, M., Waeterloos, C. and Ponnet, K. (2020) Adoption of a contact tracing app for containing Covid-19: A health belief model approach. *JMIR Public Health and Surveillance* 6 (3), e20572.

Wang, C.J., Ng, C.Y. and Brook, R.H. (2020) Response to Covid-19 in Taiwan: Big data analytics, new technology, and proactive testing. *Journal of the American Medical Association* 323 (14), 1341–1342.

Wang, D. (2016) *Renfang Research on Urban Housing Lease-Purchase Choose Factors*. Nanjing: Nanjing Tech University.

Wang, D., Xiang, Z. and Fesenmaier, D.R. (2014) Adapting to the mobile world: A model of smartphone use. *Annals of Tourism Research* 48, 11–26.

Wang, D., Xiang, Z., Law, R. and Ki, T.P. (2016) Assessing hotel-related smartphone apps using online reviews. *Journal of Hospitality Marketing & Management* 25 (3), 291–313.

Wang, N. (2000) *Tourism and Modernity. A Sociological Analysis*. Oxford: Pergamon Press.

Watanabe, T. and Omori, Y. (2020) *Online Consumption During and After the Covid-19 Pandemic: Evidence from Japan.* Tokyo: CREPE.

Wattpad (2019) The joy of missing out. How Gen Z is finding balance in an upside-down world. See https://brands.wattpad.com/gen-z-report-jomo (accessed 3 September 2020).

Watts, R. (2008) A gendered journey: Travel of ideas in England c. 1750–1800. *History of Education* 37 (4), 513–530.

Wearing, B. (1990) Beyond the ideology of motherhood: Leisure as resistance. *Australian and New Zealand Journal of Sociology* 26 (1), 36–58.

Wearing, B. (1998) *Leisure and Feminist Theory.* London: Sage.

Wearing, B. and Wearing, S. (1988) 'All in a day's leisure': Gender and the concept of leisure. *Leisure Studies* 7 (2), 111–123.

Weaver, A. (2011) The fragmentation of markets, neo-tribes, nostalgia, and the culture of celebrity: The rise of themed cruises. *Journal of Hospitality and Tourism Management* 18 (1), 54–60.

Wee, D. (2019) Generation Z talking: Transformative experience in educational travel. *Journal of Tourism Futures* 5 (2), 157–167.

Weeden, C., Lester, J.A. and Thyne, M. (2011) Cruise tourism: Emerging issues and implications for a maturing industry. *Journal of Hospitality and Tourism Management* 18 (1), 26–29.

West, C. and Zimmerman, D-H. (1987) Doing gender. *Gender and Society* 1 (2), 125–151.

Westermarck, E. (1908) *The Origin and Development of the Moral Ideas.* London: Palgrave MacMillan.

Westwood, S., Pritchard, A. and Morgan, N.J. (2000) Gender-blind marketing: Businesswomen's perceptions of airline services. *Tourism Management* 21 (4), 353–362.

White, R. (2005) *On Holidays: A History of Getting Away in Australia.* North Melbourne: Pluto Press.

Whitmore, A., Agarwal, A. and Xu, L.D. (2015) The internet of things: A survey of topics and trends. *Information Systems Frontiers* 17, 261–274.

Whyte, L.B. and Shaw, S.M. (1994) Women's leisure: An exploratory study of fear of violence as a leisure constraint. *Journal of Applied Recreation Research* 19 (1), 5–21

Williams, K.C. and Page, R.A. (2011) Marketing to the generations. *Journal of Behavioural Studies in Business* 3 (3), 1–17.

Wilson, E. and Little, D.E. (2005) A 'relative escape'? The impact of constraints on women who travel solo. *Tourism Review International* 9, 155–175.

Wilson, E. and Harris, C. (2006) Meaningful travel: Women, independent travel and the search for self and meaning. *Tourism* 54 (2), 161–172.

Wilson, E. and Little, D.E. (2008) The solo female travel experience: Exploring the 'geography of women's fear'. *Current Issues in Tourism* 11 (2), 167–186.

Wohl, R. (1979) *The Generation of 1914.* Cambridge: Harvard University Press.

Wood, S. (2013) Generation Z as consumers: Trends and innovation. See https://iei.ncsu.edu/wp-content/uploads/2013/01/GenZConsumers.pdf (accessed 18 March 2020).

Woodman, D. (2016) The sociology of generations and youth studies. In A. Furlong (ed.) *Routledge Handbook of Youth and Young Adulthood* (pp. 36–42). Routledge Handbook Online.

Woodman, D. (2018) Using the concept of generation in youth sociology. In A. Lange, H. Reiter, S. Schutter and C. Steiner (eds) *Handbuch Kindheits- und Jugendsoziologie* (pp. 97–107). Springer Reference Sozialwissenschaften. Wiesbaden: Springer VS.

Woyo, E. (2021) The sustainability of using domestic tourism as a post-Covid-19 recovery strategy in a distressed destination. *Information and Communication Technologies in Tourism* 1, 476–489.

Wu, S.I., Wei, P.L. and Chen, J.H. (2008) Influential factors and relational structure of internet banner advertising in the tourism industry. *Tourism Management* 29 (2), 221–236.

Wuest, B., Welkey, S., Mogab, J. and Nicols, K. (2008) Exploring consumer shopping preferences: Three generations. *Journal of Family and Consumer Sciences* 100 (1), 31–37.

Wunderman Thompson Commerce (2019) Generation alpha. Preparing for the future consumer 2019. See https://insights.wundermanthompsoncommerce.com/generation-alpha-2019 (accessed 1 September 2020).

Wyn, J. and Woodman, D. (2006) Generation, youth and social change in Australia. *Journal of Youth Studies* 9 (5), 37–41.

Wyse Travel Confederation (2018) New horizons survey IV: A global study of the youth and student traveller. See https://www.wysetc.org/wp-content/uploads/sites/19/2018/06/New-Horizons-IV_Preview.pdf (accessed 18 March 2020).

Xiang, Z. and Fesenmaier, D.R. (2017) *Analytics in Smart Tourism Design. Concepts and Methods*. Berlin: Springer.

Xiang, Z., Wang, D., O'Leary, J.T. and Fesenmaier, D.R. (2015) Adapting to the internet trends in travellers: Use of the web for trip planning. *Journal of Travel Research* 54 (4), 511–527.

Yang, S.B. and Guy, M.E. (2006) GenXers versus boomers: Work motivators and management implications. *Public Performance and Management Review* 29 (3), 267–284.

Yates, A., Starkey, L., Egerton, B. and Flueggen, F. (2020) High school students' experience of online learning during Covid-19: The influence of technology and pedagogy. *Technology, Pedagogy and Education* 30 (1), 59–73.

Yelkur, R. (2003) A comparison of buyer behaviour characteristics of U.S. and French generation X. *Journal of Euromarketing* 12 (1), 5–17.

Yeoman, I. (2008) *Tomorrow's Tourist: Scenarios & Trends*. Oxford: Butterworth Heinemann.

Yeoman, I. (2012) *2050 – Tomorrow's Tourism*. Bristol: Channel View Publications.

Yeoman, I. and McMahon-Beattie, U. (eds) (2020) *The Future Past of Tourism: Historical Perspectives and Future Evolutions*. Bristol: Channel View Publications.

Zimdars, M. and McLeod, K. (2020) *Fake News: Understanding Media and Misinformation in the Digital Age*. Cambridge, MA: MIT Press.

Zinola, A. (2018) *La rivoluzione nel carrello*. Milan: Guerini Next.

Žižek, S. (2020) *Pandemic! Covid-19 Shakes the World*. New York: OR Books.

Zukin, S. (1995) *The Cultures of Cities*. Cambridge: Blackwell.

Index

Africa: 56, 65, 90, 102n8, 104, 105, 110
Airbnb: 29, 40, 71, 77, 79, 81
Amazon: 31, 43, 48, 71
America: 30–32, 56, 65, 68, 102n8, 104–106, 109
Annals of Tourism Research: 85
Artificial Intelligence (AI): *ix*, 43, 48, 49, 51, 60, 77, 97
Asia (Asia Pacific): 30–32, 56, 65, 69, 75, 90, 102n8, 105
Australia: 14, 25n3, 40, 97, 102n6, 104
Authenticity: 22, 60, 77, 78, 80

Baby Boomer Generation (Baby Boomers): *x*, 11, 12, 14–17, 19, 33, 54–56, 71, 93, 94, 114, 116, 119
Bontekoning, Aart: 13
Booking (Booking.com): 36, 39
Bourdieu, Pierre: 12
Brazil: 25n3, 69, 97, 98, 104

Canada: 14, 25n3, 75, 105, 114
Car Sharing: 68–70, 74
China: 14, 24, 25n3, 27–30, 33–35, 38, 40, 56, 58, 65–67, 69, 70, 73–75, 78, 98, 102n6, 110
Climate Change: *xi*, 18, 21, 28, 39, 58, 59, 99, 124
Colombia: 39
Communitycation: 42
Comte, August: 5, 10
Covid-19, *viii*, *xi*, 28, 30, 31, 40, 61, 81, 87, 98–101, 102n8, 112, 121, 122, 124

Denmark: 89, 91
Digital Technologies: *x*, 33, 50, 60, 121
Dilthey, Wilhelm: 5, 9
Discrimination: *xi*, 80, 99, 106, 108, 114, 117, 118

East (Eastern)
    Countries: 24
    Markets: 32
    World: 27, 72
Eco-anxiety: 39
E-commerce: 30–33, 66, 73
Eisenstadt, Shmuel: 4, 5
Environment(Environmental, Enviromentally)
    Issues: *ix*, *xi*, 4, 16–19, 21, 22, 24, 26, 28, 32, 33, 35, 38–41, 43, 48, 50, 54, 57, 58, 69, 73, 74, 79, 82, 90, 97, 99, 111, 112, 114, 115, 120, 122–124
Europe: 6, 23, 31, 32, 36, 56, 61, 65, 69, 72, 74, 75, 80, 89–91, 102n8, 104, 105, 107, 108, 123
Eyerman, Ron: 9, 12

Facebook: 33, 34, 36, 42, 48, 49, 56, 61
Feminism: 92, 95
Food: 29–31, 40, 67, 100, 115
France: 89, 91, 98, 106
Future
    Challenges: *vii*, 12, 124
    of Tourism: 98, 114, 117
    Travel Behaviour: 14, 59

Gay (Gay Men): *xi*, 19, 94, 95, 103, 104, 106–111, 114–118

Gender
  Ideologies: 86, 88, 92, 98
  Inequalities: *xi*, 87, 94, 99, 120
  Perspective: *xi*, 85, 86, 108, 118
  Roles: *xi*, 87, 93, 95, 96
  Stereotypes: 27, 86, 92, 93, 98
  and Tourism: 85
Generational
  Perspective: *vii*, *viii*, 24, 120
  Theories: *viii*, 4–6, 10, 22
  Turnover: *vii*, *viii*, *xi*, 12, 103
  Values: 53
Generation Alpha (Gen Alpha): *viii*, *ix*, 15, 22, 97, 111, 125
Generation X: 11, 12, 14, 15, 17, 18, 21, 30, 33, 71
Generation Y: 14, 66, 111
Generation Z (Gen Zers): *viii*–*x*, 15, 20–23, 26, 28, 29, 31–34, 36–39, 42, 50, 57–60, 85, 93–96, 99, 102n8, 103, 105, 112, 114, 115, 122, 124, 125
Germany: 25n3, 58, 80, 89, 98, 105
Globalisation: 12, 13, 18, 19, 26
Global Warming: *xi*, 28
Greatest Generation: 14, 15

Historical Events: *vii*, 3, 9, 10, 13, 23
Homosexual People (Homosexual Community): 103, 104, 106, 108–110, 114, 117
Homosexual Tourism: 104, 106–108, 110
Howe, Neil, *viii*: 10, 11, 13–16, 19

India: 29, 30, 56, 69, 97, 98, 102n6
Indonesia: 30, 56, 104
Influencer: 33–35, 37, 38
Instagram
  Instagrammability: 36–38
  Instagrammable: 37, 58
  Instagram Story: 34
Italy: 39, 70, 71, 80, 89, 98, 105, 107

Jamaica: 110
Japan: 24, 25n3, 104

Kinship (Kinship Relations): 4, 5

Latin America: 56, 65, 102n8, 104
Leisure: *xi*, 15–17, 20, 25n3, 30, 48, 49, 53, 67, 85–89, 97, 102n5, 102n6, 103, 110, 111, 117, 118, 121
Lesbian (Lesbian women): *xi*, 19, 94, 95, 103, 104, 106, 110, 111, 115–117
LGBT
  LGBTQ+: *xi*, 103–119
  LGBTQ+ Community: 115–117
  LGBTQ+ Tourism: 105, 108–112
  Rights: 20, 95
LinkedIn: 34
Longitudinal Research (Longitudinal Study): 24, 59

Mannheim, Karl: *vii*, *viii*, *x*, 3, 4, 6–12, 24, 93
Marriage: *xi*, 19, 20, 27, 28, 55, 65, 86, 89, 95, 104, 105
Masculinity: 88, 96
Mentré, François: 6, 10
Mexico: 25n3, 69
Middle East: 65, 90, 102n8, 104, 105
Mill, John Stuart: 5, 10
Millennial Generation (Millennials): 8, 15, 18, 19, 54, 57, 122

North America: 10, 56, 65, 68, 102n8, 104–106, 108, 109
Norway: 80, 90, 91

Ortega y Gasset, José: 6, 8, 10, 13

Pakistan: 30, 56
Pew Research Center: 19–21, 23, 93–95, 102n2, 124
Pinder, Wilhelm: 6, 24n1

QQ (QZone): 35

Rainbow Tourism (Rainbow Touristification): 105, 118
Reel: 29, 42, 107, 111
Resilience (Resilient): 18, 28, 87, 99, 101, 112, 121, 124
Rifkin, Jeremy: 68
Robot: 48, 97, 115
Russia: 69, 104, 105, 110

Sharing (Sharing Economy): x, 29, 30, 63, 71, 78–81, 97
Silent Generation: 11, 14–17, 19, 93, 94, 96
Smartphone: ix, 21, 36, 37, 43, 47, 48, 51, 56, 57, 67, 122
Snapchat: 34
Social Change: vii, ix, 3–5, 9, 15, 16, 26, 86, 108, 112
Social Network: viii, x, 29, 30, 33–38, 41, 50, 56, 57, 60, 62, 64, 68, 80, 81, 98, 116, 118
Social Sciences: 4, 13, 63, 104
Solo Travel: 20, 87, 92, 98
Sorokin, Piotr: 4
Spartacus International Gay Guide: 116
Strauss, William: viii, 10, 11, 13–16, 19
Sustainability: ix, 20–22, 38, 57, 73, 78, 101, 124
Sustainable Development: ix, 82
Sweden: 89, 91, 104

Taiwan: 24
Taobao: 30, 31, 73
Technological Device: 28, 33, 50, 55, 59, 74, 120
Technological Innovation (Technological Advances): vii, 15, 21, 50, 98, 121
Telegram: 42
Thunberg, Greta: 28
TikTok (TikToker): 29, 32, 34, 48, 61
Tourism
    Choices: viii, x, 4, 97, 114, 120
    Experiences: 51, 59–62, 78, 85, 110
    Preferences: 13–22
Tourist Gaze: 40, 60
Transgender People (Transgender Persons): xi, 19, 94, 95, 103, 108, 109, 116, 118
Triller: 29

Turner, Bryan S., 9, 12, 13, 16, 94
Twitch: 29
Twitter: 34, 49, 56

Uber: 71, 78, 108
Ukraine: 104, 105
United Kingdom (UK): 25n3, 49, 58, 69, 80, 97, 98, 102n6, 105, 117
United States (US): viii, 14, 15, 23, 25n3, 39, 58, 69, 72, 75, 90, 95, 97, 98, 102n5, 102n6, 105, 108, 111
Urban Areas (Urban Spaces): 69, 74, 88, 110
Urbanisation: 27, 90

Vegan (Veganism): 31, 40
Vegetarian (Vegetarianism): 30, 31
Violence
    gender based violence (GBV): 95, 101
    against LGBTQ+ people: 116
    against women: 99
Virtual Reality (VR): 43, 48, 49, 51, 60

WeChat (Wechatability): 35, 38, 74
Weibo: 35, 36
West (Western)
    Countries: 23, 91
    Markets: 58
    World: 17, 27, 29, 33, 65, 66, 90
WhatsApp: 42, 55
Williams Institute: 112
Women's Employment: 18, 100
Women's Travel (Women's Tourism): 18, 87, 100

Young Travellers: 22, 27, 36, 40, 41, 50, 61, 62, 73, 74, 115
YouTube (Youtuber): 33, 34, 49

ZiRoom: 66, 67

ty Concerns and Information please contact our EU Authorised

System Europe

e tee 50

Tallinn

nia

psr.requests@easproject.com

www.ingramcontent.com/pod-product-compliance
Ingram Content Group UK Ltd.
Pitfield, Milton Keynes, MK11 3LW, UK
UKHW021943200326

1879IPUK00004B/65